Nuclear Magnetic Resonance

P. J. Hore

Reader in Physical Chemistry and Fellow of Corpus Christi College, Oxford

Series sponsor: **ZENECA**

ZENECA is a major international company active in four main areas of business: Pharmaceuticals, Agrochemicals and Seeds, Specialty Chemicals, and Biological Products.

ZENECA's skill and innovative ideas in organic chemistry and bioscience create products and services which improve the world's health, nutrition, environment, and quality of life.

ZENECA is committed to the support of education in chemistry and chemical engineering.

OXFORD

UNIVERSITY PRESS

OXFORD

UNIVERSITY PRESS

Great Clarendon Street, Oxford OX2 6DP
Oxford University Press is a department of the University of Oxford.
It furthers the University's objective of excellence in research, scholarship,
and education by publishing worldwide in

Oxford New York

Athens Auckland Bangkok Bogotá Buenos Aires Calcutta
Cape Town Chennai Dar es Salaam Delhi Florence Hong Kong Istanbul
Karachi Kuala Lumpur Madrid Melbourne Mexico City Mumbai
Nairobi Paris São Paolo Singapore Taipei Tokyo Toronto Warsaw

with associated companies in Berlin Ibadan

Oxford is a registered trade mark of Oxford University Press
in the UK and in certain other countries

Published in the United States
by Oxford University Press Inc., New York

© P. J. Hore, 1995
First published 1995
Reprinted 1996, 1998, 2000

A catalogue record for this book is available from the British Library

Library of Congress Cataloging in Publication Data
Hore, P. J.
Nuclear magnetic resonance P. J. Hore
(Oxford chemistry primers: 32)
Includes bibliographical references and index.
1. Nuclear magnetic resonance spectroscopy.   I. Title   II. Series
QD96.N8H67   1995   543.0877—dc20   94-46220

ISBN 0 19 855682 9 (Pbk)

Printed in Great Britain by
The Bath Press, Bath

Series Editor's Foreword

Oxford Chemistry Primers are designed to provide clear and concise introductions to a wide range of topics that may be encountered by chemistry students as they progress from the freshman stage through to graduation. The Physical Chemistry series will contain books easily recognized as relating to established fundamental core material that all chemists will need to know, as well as books reflecting new directions and research trends in the subject, thereby anticipating (and perhaps encouraging) the evolution of modern undergraduate courses.

In this Physical Chemistry Primer, Peter Hore presents a clearly written and elegant introductory account of *Nuclear Magnetic Resonance*—a spectroscopic technique of unrivaled power and versatility. The book explains in simple terms those basic ideas and applications of NMR which are now essential for any practising chemists. This Primer will be of interest to all students of chemistry (and their mentors).

Richard G. Compton
Physical Chemistry Laboratory, University of Oxford

Preface

Nuclear magnetic resonance is an enormously powerful and versatile technique for investigating the structure and dynamics of molecules. These days it is difficult to find a Chemistry laboratory without an NMR spectrometer or an undergraduate Chemistry course without a set of NMR lectures. This book offers a clear, concise introduction to the physical principles of NMR of liquids, and the interactions that determine the appearance of NMR spectra. The six chapters describe and explain how nuclear spins interact with a magnetic field (the chemical shift) and with each other (spin–spin coupling); how NMR spectra are affected by chemical equilibria (exchange) and molecular motion (relaxation); and concludes with an outline of the workings of a few one- and two-dimensional NMR experiments. I have made every effort to keep things simple: only essential mathematics and theory are included. The emphasis throughout is on understanding the foundations of the technique and how it may be used to study problems of chemical interest.

The shape and content of this short book owe much to those who taught me magnetic resonance—Keith McLauchlan, Rob Kaptein and Ray Freeman—especially to Keith and Ray whose Oxford undergraduate lecture courses I inherited ten years ago, and on which the book is based. I am indebted and profoundly grateful to all three. My thanks also go to Jonathan Jones, who carefully and perceptively read the whole thing and made penetrating comments on almost every page; to Paul Hodgkinson, Mark Wormald and Pete Biggs who cheerfully untangled the computer glitches I encountered (and sometimes generated) while drawing the figures; and to Craig Morton and Mark Wormald for generously making available their beautiful spectra.

Oxford
October 1994   P. J. H.

Contents

1 Introduction

Molecules are inconveniently small—too small to be observed and studied directly. One therefore needs a spy, capable of relaying information on the structures, motions, and chemical reactions of molecules without significantly modifying those properties, and versatile enough to report on a wide range of molecules in a variety of situations. The spies that form the subject of this book are atomic nuclei, and the attribute that makes them so successful at espionage is their magnetism.

Nuclear magnetic moments are exquisitely sensitive to their surroundings and yet interact very weakly with them. Most elements have at least one naturally occurring magnetic isotope, so essentially every molecule one might wish to investigate has one or more spies already in place, exerting a negligible influence on the molecular properties they probe.

When a magnetic nucleus is placed in a magnetic field, it adopts one of a small number of allowed *orientations* of different energy. For example, the nucleus of the hydrogen atom (a proton) has just two permitted orientations, as shown in Fig. 1.1. Roughly speaking, the magnetic moment can point in the *same* direction as the field or in the *opposite* direction. These two states are separated by an energy ΔE, which depends, not surprisingly, on the strength of the interaction between the nucleus and the field, i.e. on the size of the nuclear magnetic moment and the strength of the magnetic field. ΔE may be measured by applying electromagnetic radiation of frequency ν, which causes nuclei to 'flip' from the lower energy level to the upper one, provided the *resonance condition* $\Delta E = h\nu$ (h is Planck's constant) is satisfied. This is *nuclear magnetic resonance* (or NMR) spectroscopy.

For a given field strength, the energy gap ΔE and therefore the resonance frequency, are determined principally by the nuclide observed, because every nuclide (e.g. ^{1}H, ^{2}H, ^{13}C, ^{14}N, ^{15}N, etc.) has a characteristic magnetic moment. But there is far more to NMR than being able to distinguish hydrogen from deuterium, or ^{13}C from ^{14}N. Fortunately for chemists, the resonance frequency also depends, slightly, on the chemical environment of the nucleus in a molecule, an effect known as the *chemical shift*. For example, Fig. 1.2 shows an NMR spectrum of the hydrogen nuclei in liquid ethanol, CH_3CH_2OH. The three different kinds of H atom have different resonance frequencies and so give rise to three separate peaks, identifiable by their integrated areas which are in the ratio 1 : 2 : 3 reflecting the number of protons of each type: OH, CH_2, CH_3. This rather low quality spectrum, recorded in 1951, hints at the way in which NMR can be used to probe individual atoms within molecules. As we shall see, NMR is an enormously powerful technique for studying molecules at the atomic level.

I am grateful to Prof. R. Freeman for the spy simile.

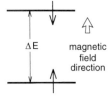

Fig. 1.1 Energy levels of a hydrogen nucleus in a magnetic field. The two permitted orientations of the nuclear magnetic moment relative to the magnetic field direction have energies differing by ΔE.

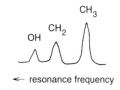

Fig. 1.2 ^{1}H NMR spectrum of liquid ethanol, CH_3CH_2OH. (Adapted from J. T. Arnold, S. S. Dharmatti and M. E. Packard, *J. Chem. Phys.*, 1951, **19**, 507.)

1.1   Angular momentum and nuclear magnetism

Having seen very briefly what NMR is, let us go back to the beginning and look at the origins and properties of nuclear magnetism.

Spin angular momentum

Magnetic nuclei possess an intrinsic angular momentum known as *spin*, whose magnitude is quantized in units of $\hbar (=h/2\pi)$:

$$\text{magnitude of spin angular momentum} = [I(I+1)]^{\frac{1}{2}}\hbar. \qquad (1.1)$$

The *spin quantum number I* of a nucleus may have one of the following values

$$I = 0, \tfrac{1}{2}, 1, \tfrac{3}{2}, 2, \ldots, \qquad (1.2)$$

with quantum numbers greater than 4 being rather rare. Spin quantum numbers for a selection of nuclei are given in Table 1.1. Notice that isotopes of the same element may have different I and that some common nuclei, notably ^{12}C and ^{16}O, have $I = 0$, i.e. no angular momentum, no magnetic moment and consequently no NMR spectra. Protons, neutrons, and electrons all have $I = \tfrac{1}{2}$; they are said to be 'spin-$\tfrac{1}{2}$' particles.

The spin quantum number of a nucleus is determined largely by the number of unpaired protons and neutrons. For example, an isotope such as ^{12}C has even numbers of protons and neutrons: all the protons pair up with antiparallel spins, as do all the neutrons, giving a net spin angular momentum of zero ($I = 0$). A nucleus with odd numbers of protons and neutrons (e.g. ^{14}N) generally has an integral, non-zero quantum number because the total number of unpaired nucleons is even, and each of them contributes $\tfrac{1}{2}$ to the quantum number. However, it is difficult to predict exactly how many protons and neutrons will be unpaired except in cases such as ^{2}H. These ideas can be extended to nuclei with even numbers of protons and odd numbers of neutrons, or vice versa, which usually have a half-integral quantum number due to an odd number of unpaired nucleons. These rules of thumb, which are not infallible, are collected in Table 1.2.

Table 1.1 Nuclear spin quantum numbers (*I*) of some commonly occurring nuclides

I	Nuclide
0	^{12}C, ^{16}O
$\tfrac{1}{2}$	^{1}H, ^{13}C, ^{15}N, ^{19}F, ^{29}Si, ^{31}P
1	^{2}H, ^{14}N
$\tfrac{3}{2}$	^{11}B, ^{23}Na, ^{35}Cl, ^{37}Cl
$\tfrac{5}{2}$	^{17}O, ^{27}Al
3	^{10}B

Table 1.2 Rules for predicting nuclear spin quantum numbers (*I*) from the numbers of protons and neutrons in a nucleus

Number of protons	Number of neutrons	I
even	even	0
odd	odd	1 or 2 or 3 or . . .
even	odd	$\tfrac{1}{2}$ or $\tfrac{3}{2}$ or $\tfrac{5}{2}$ or . . .
odd	even	$\tfrac{1}{2}$ or $\tfrac{3}{2}$ or $\tfrac{5}{2}$ or . . .

Space quantization

Spin angular momentum is a vector quantity; its direction, as well as its magnitude, is quantized. The angular momentum $\boldsymbol{I}$ (vectors are printed in bold-face type; I should not be confused with the quantum number I) of a spin-I nucleus has just $2I+1$ projections onto an arbitrarily chosen axis, say the z axis. That is, the z component of I, denoted I_z, is quantized:

$$I_z = m\hbar \tag{1.3}$$

where m, the magnetic quantum number, has $2I+1$ values in integral steps between $+I$ and $-I$:

$$m = I, I-1, I-2, \ldots -I+1, -I. \tag{1.4}$$

For example, the angular momentum of a spin-$\frac{1}{2}$ nucleus (e.g. ^{1}H, ^{13}C) has two permitted directions, $I_z = \pm\frac{1}{2}\hbar$, while a nucleus with $I = 1$ has three possible states, $I_z = 0, \pm\hbar$. This *space quantization* is illustrated in Fig. 1.3. Remember that the axis of quantization is arbitrary: in the absence of a magnetic field, the spin angular momentum has no preferred direction(s).

Nuclear magnetization

The magnetic moment of a nucleus is intimately connected with its spin angular momentum. To be more precise, the magnetic moment **μ** (also a vector quantity) is directly proportional to I with a proportionality constant γ, known as the gyromagnetic ratio:

$$\boldsymbol{\mu} = \gamma \boldsymbol{I}. \tag{1.5}$$

The gyromagnetic ratios of some common NMR nuclei are given in Table 1.3. Note that the magnetic moment of a nucleus is *not* simply the sum of the magnetic moments of the constituent protons and neutrons.

To summarize, the magnetic moment of a nucleus is parallel (or occasionally antiparallel, for nuclei with negative γ) to the spin angular momentum. The magnitude and orientation of both are quantized.

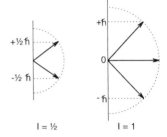

Fig. 1.3 Space quantization of spin-$\frac{1}{2}$ and spin-1 nuclei. The spin angular momentum has magnitude $[I(I+1)]^{1/2}\hbar$ and z component $m\hbar$, where m is given by eqn 1.4.

Table 1.3 Gyromagnetic ratios, NMR frequencies (in a 9.4 T field), and natural abundances of selected nuclides

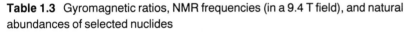

	$\gamma/10^7 \text{T}^{-1}\text{s}^{-1}$	ν/MHz	Natural abundance/%
^{1}H	26.75	400.0	99.985
^{2}H	4.11	61.4	0.015
^{13}C	6.73	100.6	1.108
^{14}N	1.93	28.9	99.63
^{15}N	−2.71	40.5	0.37
^{17}O	−3.63	54.3	0.037
^{19}F	25.18	376.5	100.0
^{29}Si	−5.32	79.6	4.70
^{31}P	10.84	162.1	100.0

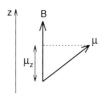

Fig. 1.4 The relationship between the magnetic field **B**, the nuclear magnetic moment μ, and its component along the field direction, μ_z (the scalar product of **B** and μ).

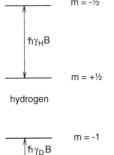

Fig. 1.5 Energy levels for hydrogen ($I = \frac{1}{2}$) and deuterium ($I = 1$) nuclei in a magnetic field **B**. Note the effect of the different gyromagnetic ratios of the two nuclei (the energy-level splittings are not drawn to scale).

Effect of a magnetic field

In the absence of a magnetic field, all $2I + 1$ orientations of a spin-I nucleus have the same energy. This degeneracy is removed when a magnetic field is applied: the energy of a magnetic moment μ in a magnetic field **B** (yet another vector) is minus the scalar product of the two vectors:

$$E = -\mu \cdot B. \tag{1.6}$$

In the presence of a strong field, the quantization axis z is no longer arbitrary, but coincides with the field direction. Therefore (Fig. 1.4):

$$E = -\mu_z B \tag{1.7}$$

where μ_z is the z component of μ (the projection of μ onto **B**) and B is the strength of the field (the magnitude of **B**). From eqns 1.3 and 1.5, $\mu_z = \gamma I_z$, and $I_z = m\hbar$; whence

$$E = -m\hbar\gamma B. \tag{1.8}$$

That is, the energy of the nucleus is shifted by an amount proportional to the magnetic field strength, to the gyromagnetic ratio and the z component of the angular momentum. The $2I + 1$ states for a spin-I nucleus are equally spaced, with energy gap $\hbar\gamma B$ (Fig. 1.5).

The selection rule for NMR is $\Delta m = \pm 1$; the allowed transitions are therefore between adjacent energy levels. The resonance condition $\Delta E = h\nu$ is thus

$$\Delta E = h\nu = \hbar\gamma B \tag{1.9}$$

or

$$\nu = \frac{\gamma B}{2\pi} \tag{1.10}$$

where ν is the frequency of the electromagnetic radiation. All $2I$ allowed transitions for a spin-I nucleus have the same energy. More information on the quantization of angular momentum may be found in Atkins (1983).

The magnetic field experienced by a nucleus *in a molecule* differs slightly from the external field such that the exact resonance frequency is characteristic of the chemical environment of the nucleus. This interaction, the chemical shift, is discussed in Chapter 2.

1.2 NMR spectroscopy

Resonance frequencies

A typical magnetic field strength used for NMR is 9.4 T (T = Tesla = 10^4 Gauss), roughly 10^5 times stronger than the earth's magnetic field. For hydrogen nuclei, eqn 1.10 predicts a resonance frequency of $\nu = 4 \times 10^8$ Hz or 400 MHz (see Table 1.3 for the value of γ). This falls in the radiofrequency region of the electromagnetic spectrum and corresponds to a wavelength of 75 cm. The radiation required to induce NMR transitions is consequently referred to as the *radiofrequency field*. Magnetic fields in the range 1.4–14.1 T are commonly used, giving proton resonance frequencies of 60–600 MHz. Since the proton is by far the most popular NMR nucleus (for reasons that will soon emerge), NMR spectrometers are usually classified by their proton frequencies, rather than the strengths of their magnetic fields. Table 1.3 summarizes the NMR properties of several magnetic nuclei: gyromagnetic ratio, resonance frequency in a 400 MHz spectrometer, and natural isotopic abundance. Of these nuclei, ^{1}H has the largest gyromagnetic ratio (and the largest magnetic moment); in fact the γ of ^{1}H is exceeded only by that of the radioactive isotope tritium, ^{3}H.

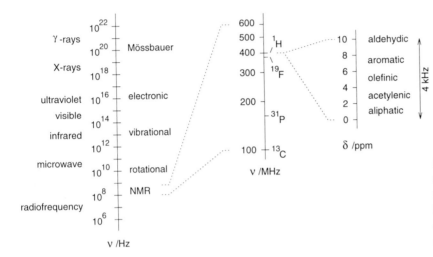

Fig. 1.6 Electromagnetic spectrum. Expanded regions show NMR frequencies of nuclei in a 9.4 T field, and typical ^{1}H chemical shifts (in parts per million, ppm).

To put NMR into context, Fig. 1.6 shows the electromagnetic spectrum, on a logarithmic scale, from the radiofrequency region ($\approx 10^6$ Hz) through microwaves, infra-red, visible, ultra-violet and X-rays to γ-rays ($\approx 10^{22}$ Hz). NMR lies at the low frequency end of the spectrum; most other spectroscopies—rotational, vibrational, electronic, Mössbauer—are concerned with larger energy level splittings and hence higher frequencies. Also shown, on an expanded logarithmic scale, is the spectral region between 100 and 600 MHz, showing the resonance frequencies of a few nuclides in a 9.4 T field. In addition, as a foretaste of the next chapter, a narrow slice of frequencies close to 400 MHz is expanded to show some typical ^{1}H chemical shifts for a few organic functional groups. Notice that the total width of this section is only 4 kHz. Although other nuclei have somewhat larger chemical shifts, the differences in resonance frequencies between nuclides almost always exceeds the chemical shift ranges, so that it is rare to find overlapping spectra from different nuclides.

Populations of energy levels

When placed in a magnetic field, a collection of magnetic nuclei spread themselves amongst the $2I + 1$ available energy levels according to the Boltzmann distribution. Let us consider, once again, protons in a 9.4 T field, at a temperature of 300 K. The ratio of populations of the two levels is

$$\frac{n_{\text{upper}}}{n_{\text{lower}}} = e^{-\Delta E/kT} \qquad (1.11)$$

where ΔE, the energy gap, equals $\hbar\gamma B$ and k is Boltzmann's constant. Evaluating this expression one finds that $\Delta E = 2.65 \times 10^{-25}$ J, $kT = 4.14 \times 10^{-21}$ J, and $\Delta E/kT = 6.4 \times 10^{-5}$. Thus, the energy required to reorient the spins is dwarfed by the thermal energy kT, so that there will be very little tendency for the spins to become ordered in the lower energy level. With such small values of $\Delta E/kT$, eqn 1.11 may be simplified using $e^{-x} \approx 1 - x$, to obtain the normalized population difference

$$\frac{n_{\text{lower}} - n_{\text{upper}}}{n_{\text{lower}} + n_{\text{upper}}} = \frac{\Delta E}{2kT}. \qquad (1.12)$$

With the above numbers, eqn 1.12 gives a population difference of 3.2×10^{-5} or one part in about 31 000. This difference will be even smaller for protons in a weaker field, or for nuclei with lower gyromagnetic ratios (i.e. almost all other nuclei). This situation is in stark contrast to electronic spectroscopy at a frequency of, say, 6×10^{16} Hz. Here ΔE is much larger than kT and almost all of the molecules will be in their ground state, leaving the excited state virtually empty.

In any form of spectroscopy, an electromagnetic field excites molecules or atoms or electrons or nuclei as the case may be, from the lower energy level to the upper one with the *same* probability as it induces the reverse transition, from excited state to ground state. The net absorption of energy, and hence the intensity of the spectroscopic transition, is therefore dependent on the *difference* in populations of the two levels. In NMR spectroscopy, where the upward transitions outnumber the downward transitions by only one in 10^4–10^6, it is as if one *detects* only one nucleus in every 10^4–10^6. Add to this the fact that spectroscopy at higher frequencies is much more sensitive as a rule, because higher energy photons are easier to detect, and it becomes clear that NMR signals must be rather weak. It is therefore of crucial importance to optimize signal strengths, for example by using strong magnetic fields to maximize ΔE. Similarly, nuclei with large gyromagnetic ratio and high natural abundance are favoured (Table 1.3): hence the popularity of ^{1}H as an NMR nucleus.

Spin–spin coupling

The chemical shift is not the only source of information encoded in an NMR spectrum. Magnetic interactions between nuclei give rise to extra NMR lines which give valuable clues to the arrangement of atoms in molecules. For example, the spectrum of ethanol in Fig. 1.2 in fact comprises eight distinct peaks (see Chapter 3) and sometimes more (Chapter 4) when recorded on a spectrometer with less severe instrumental linebroadening.

Widths of NMR lines

The NMR lines of many nuclei are exceedingly narrow: it is relatively straightforward on a modern spectrometer to resolve two resonances that differ in frequency by only one part in 10^9 (typical linewidths in ^{1}H spectra of small molecules in solution are around 0.1 Hz). This quite staggering resolution comes about because nuclear magnetic moments are feeble, and interact very weakly with their surroundings (see Chapter 5).

But not all nuclei give sharp lines. For reasons discussed in Chapter 5, the NMR lines of nuclei with spin quantum numbers greater than $\frac{1}{2}$ are often broad—100 Hz or more. The majority of NMR experiments are therefore performed on spin-$\frac{1}{2}$ nuclei, the most popular being ^{1}H, ^{13}C, ^{19}F, and ^{31}P.

Experimental methods

There are three essential requirements for an NMR experiment: a strong static magnetic field, a source of radiofrequency radiation to excite nuclei in the sample, and a method for detecting the NMR signal.

The magnetic field is most commonly provided by a superconducting solenoid—a coil of resistance-free alloy supporting a persistent current. At the time of writing, magnetic fields up to 17.6 T (750 MHz ^{1}H frequency) are commercially available.

To observe magnetic resonance, one needs coherent, monochromatic electromagnetic radiation. This is generated by passing a sinusoidally oscillating current through a coil arranged around the sample which sits inside the magnet. The electrons circulating in the coil generate an oscillating electromagnetic field whose magnetic component excites the sample, provided its frequency satisfies the resonance condition. The *coherent* radiofrequency field provokes a *coherent* response from the nuclei: it causes their magnetic moments to precess *in phase* around the strong static field at their individual resonance frequencies. This oscillating magnetization in turn induces an alternating current in a suitably placed receiver coil which is then amplified and presented as the NMR signal.

In the early days of NMR (1945–1970), experiments were carried out almost exclusively using *continuous wave* methods, with the radiofrequency field present throughout. Either one kept the electromagnetic frequency ν fixed, while sweeping the magnetic field, or vice versa, so as to bring nuclei with different chemical shifts sequentially into resonance. Since 1970, *pulse* methods have almost completely taken over. A short, intense burst of radiofrequency radiation is applied to the sample to set the nuclear moments precessing around the static field direction. This oscillating magnetization is detected as a function of time after the end of the pulse and processed in a computer to give the spectrum. As outlined in Chapter 6, pulse methods have overwhelming advantages over the continuous wave approach.

2   Chemical shifts

The NMR frequency of a nucleus in a molecule is determined principally by its gyromagnetic ratio γ and the strength B of the magnetic field it experiences (eqn 1.10):

$$\nu = \frac{\gamma B}{2\pi} \ .$$

(2.1)

Thus, protons and carbon-13 nuclei resonate respectively at 400 and 100.6 MHz in a 9.4 T field. But not all protons, nor all ^{13}C nuclei, have identical resonance frequencies: ν depends (slightly) on the position of the nucleus in the molecule, or to be more precise, on the local electron distribution. This effect, the *chemical shift*, is what makes NMR so attractive to chemists. It allows one to distinguish, for example, the three types of hydrogen atom in ethanol (Fig. 1.2) and, at the other extreme, gives separately detectable signals for the hundreds of protons in a protein (Fig. 2.1).

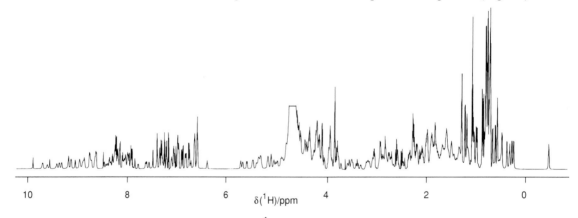

Fig. 2.1 750 MHz ^{1}H spectrum of the SH3 domain of the tyrosine kinase, Fyn, in H_2O. This protein domain has 69 amino acid residues and a molecular weight of 7688. The resonance of the solvent protons at 4.7 ppm has been truncated. This spectrum was kindly provided by C. J. Morton and J. Boyd.

2.1   Nuclear shielding

Chemical shifts arise because the field, B, actually experienced by a nucleus in an atom or molecule differs slightly from the external field B_0, i.e. the field that would be felt by a bare nucleus, stripped of its electrons. In an atom, B is slightly smaller than B_0 because the external field causes the electrons to circulate within their atomic orbitals; this induced motion, much like an electric current passing through a coil of wire, generates a small magnetic field B' in the *opposite* direction to B_0 (Fig. 2.2). The nucleus is thus *shielded* from the external field by its surrounding electrons ($B = B_0 - B'$).

B' is proportional to B_0 (the stronger the external field, the more it 'stirs up' the electrons) and typically 10^4–10^5 times smaller. Thus, the field at the nucleus may be written

$$B = B_0 - B' = B_0(1 - \sigma) \qquad (2.2)$$

where σ, the constant of proportionality between B' and B_0, is called the *shielding constant* or *screening constant*. As a result of nuclear shielding, the resonance condition (eqn 2.1) becomes

$$\nu = \frac{\gamma B_0(1 - \sigma)}{2\pi} , \qquad (2.3)$$

i.e. the resonance frequency of a nucleus in an atom is slightly lower than that of a bare nucleus, stripped of all its electrons (Fig. 2.3).

Similar effects occur for nuclei in molecules, except that the motion of the electrons is rather more complicated than in atoms and the induced fields may either *augment* or *oppose* the external field. Nevertheless, the effect is still referred to as nuclear shielding. Both the size and sign of the shielding constant in eqn 2.3 are determined by the electronic structure of the molecule in the vicinity of the nucleus. The resonance frequency of a nucleus is therefore characteristic of its environment.

Measuring chemical shifts

The shielding constant σ is an inconvenient measure of the chemical shift. Since absolute shifts are rarely needed and difficult to determine, it is common practice to define the chemical shift in terms of the *difference* in resonance frequencies between the nucleus of interest (ν) and a reference nucleus (ν_{ref}), by means of a dimensionless parameter δ:

$$\delta = 10^6 \frac{(\nu - \nu_{ref})}{\nu_{ref}} . \qquad (2.4)$$

The frequency difference $\nu - \nu_{ref}$ is divided by ν_{ref} so that δ is a molecular property, independent of the magnetic field used to measure it. The factor of 10^6 simply scales the numerical value of δ to a more convenient size: δ values are quoted in *parts per million*, or ppm.

To see how δ is related to σ, eqns 2.3 and 2.4 can be combined to give

$$\delta = 10^6 \frac{\sigma_{ref} - \sigma}{1 - \sigma_{ref}} \approx 10^6(\sigma_{ref} - \sigma) \qquad (2.5)$$

where $\sigma_{ref} \ll 1$ has been used. An increase in σ (greater shielding) leads to a decrease in δ: δ is thus a *de*shielding parameter.

The reference signal is most conveniently obtained by adding a small amount of a suitable compound to the NMR sample. For ^1H and ^{13}C spectra this is usually tetramethylsilane $(CH_3)_4Si$, or TMS. This molecule is inert, soluble in most organic solvents, and gives a single, strong ^1H resonance from its 12 identical protons. Moreover, both ^1H and ^{13}C nuclei are

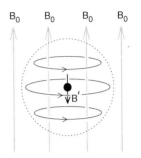

Fig. 2.2 An applied magnetic field B_0 causes the electrons in an atom to circulate within their orbitals. This motion generates an extra field B' at the nucleus in opposition to B_0.

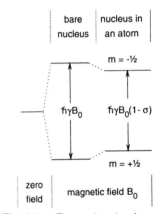

Fig. 2.3 Energy levels of a spin-$\frac{1}{2}$ nucleus.

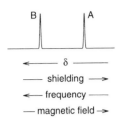

Fig. 2.4 NMR spectra are conventionally plotted with chemical shift δ increasing from right to left. Nucleus A, which is more strongly shielded than B, thus appears to the right of B, has a lower resonance frequency than B, and is sometimes referred to as being 'upfield' of B.

quite strongly shielded (large σ), so that the TMS resonance falls at the low frequency end of the spectrum: most ^{1}H and ^{13}C chemical shifts are therefore conveniently positive (δ > 0).

Conventionally, NMR spectra are plotted with δ (and ν) increasing from right to left. Thus, more heavily shielded nuclei (larger σ, smaller ν, smaller δ) appear towards the right-hand side of the spectrum. Occasionally, chemical shifts are referred to as 'upfield' or 'downfield', meaning 'more shielded' or 'less shielded' respectively. These terms date from the early days of NMR when spectra were recorded by sweeping the magnetic field strength B_0, keeping the radiofrequency fixed: the more shielded nuclei (larger σ) required a larger B_0 to come into resonance (see eqn 2.3). Although field-swept NMR is rare now, these terms linger on. The various conventions are summarized in Fig. 2.4.

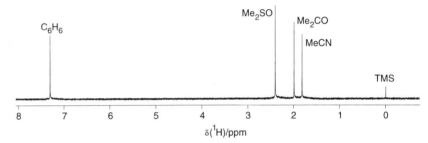

Fig. 2.5 400 MHz ^{1}H NMR spectrum of a mixture of benzene, dimethylsulphoxide, acetone, acetonitrile and tetramethylsilane.

As a simple example, Fig. 2.5 shows the 400 MHz ^{1}H spectrum of a mixture of compounds with a small amount of TMS added. Each of the five molecules has a single group of identical protons, and hence a single chemical shift. Notice that the chemical shift scale covers about 10 ppm, a typical range for protons. Chemical shifts can easily be converted back into frequencies using eqn 2.4: for example, the acetone peak in Fig. 2.5 has a δ of 2.0 ppm, so that

$$\nu_{\text{acetone}} - \nu_{\text{TMS}} = (\delta/\text{ppm})(\nu_{\text{TMS}}/10^6) = 2.0 \times 400 \text{ Hz} = 800 \text{ Hz}. \quad (2.6)$$

On a 100 MHz spectrometer, the chemical shift of acetone is still 2.0 ppm, but the resonance frequency relative to TMS is reduced in proportion to B_0 to 200 Hz.

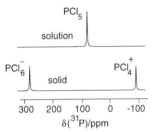

Fig. 2.6 ^{31}P NMR spectra of phosphorus pentachloride in the solid state and in solution in CS_2. The former was obtained using a technique known as *magic angle spinning* to remove the large linebroadening caused by, amongst other things, the dipolar interactions between nuclei in the solid. Chemical shifts are quoted relative to an 85% aqueous solution of orthophosphoric acid (δ = 0). (Adapted from E. R. Andrew, *Phil. Trans. R. Soc.*, 1981, **A299**, 505.)

Examples

As we shall see later, so many factors play a role in determining the size of chemical shifts that it is often impossible to relate experimental measurements *quantitatively* to molecular structure. However, valuable information can readily be extracted from NMR spectra simply by noting the number of resonances and their relative intensities, as the simple examples in the following paragraphs illustrate.

In CS_2 solution, phosphorus pentachloride, PCl_5, has a single ^{31}P resonance (Fig. 2.6), as might be expected from the trigonal bipyramidal

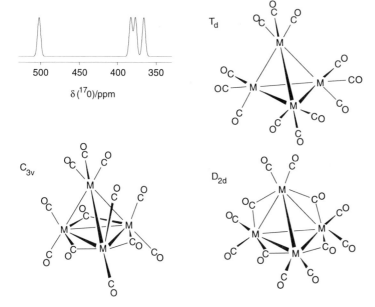

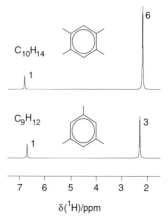

Fig. 2.7 The ^{17}O NMR spectrum of $Co_4(CO)_{12}$ in chloroform at $-25°C$ is consistent with the C_{3v} structure shown, but not the T_d or D_{2d} forms.

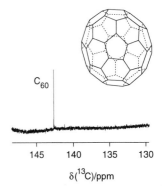

Fig. 2.8 ^{1}H spectra of compounds with molecular formulae C_9H_{12} and $C_{10}H_{14}$. The relative intensities of the peaks, obtained by integration, are as shown.

structure found in the gas phase. Solid phosphorus pentachloride, however, has *two* equally intense ^{31}P peaks, revealing clearly that the change of phase is accompanied by a change in structure (actually the disproportionation $2PCl_5 \rightarrow PCl_6^- + PCl_4^+$).

The ^{17}O NMR spectrum of $Co_4(CO)_{12}$ in chloroform at low temperature comprises four equally intense lines (Fig. 2.7), consistent with a bridged structure containing four distinct types of carbonyl. This spectrum clearly rules out a non-bridged $Ir_4(CO)_{12}$-type structure (T_d symmetry), in which all 12 carbonyls are in identical environments. It also provides strong evidence against the D_{2d} structure at one time proposed for $Co_4(CO)_{12}$ in which there are *three* distinct carbonyl environments: one bridging and two terminal.

Occasionally, *unambiguous* molecular structures can be deduced solely from the relative intensities of NMR lines. For example, the only possible structures for the compounds C_9H_{12} and $C_{10}H_{14}$ with the ^{1}H spectra shown in Fig. 2.8 are 1,3,5-trimethylbenzene and 1,2,5,6-tetramethylbenzene, respectively.

A dramatic illustration of the use of chemical shifts is provided by the ^{13}C NMR spectrum of the fullerene C_{60} (Fig. 2.9). The observation of a single NMR line for this unusual molecule provides direct evidence for its highly symmetrical football-like structure in which all 60 carbon atoms are in identical environments.

Zeolites are aluminosilicates built from corner-sharing SiO_4 and AlO_4 tetrahedra. Each silicate and aluminate group is linked, via oxygen bridges, to four other tetrahedra to give framework structures containing cavities and channels, which confer useful catalytic properties. Up to five distinct chemical shifts can be observed in the ^{29}Si NMR spectra of powdered zeolites (Fig. 2.10), corresponding to Si atoms linked to n AlO_4 and $(4-n)$

Fig. 2.9 ^{13}C NMR spectrum of buckminsterfullerene, C_{60}. (Adapted from R. Taylor, J. P. Hare, A. K. Abdul-Sada, and H. W. Kroto, *Chem. Commun.*, 1990, 1423.)

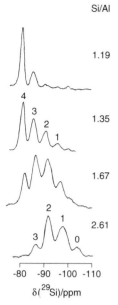

Si/Al

1.19

1.35

1.67

2.61

-80 -90 -100 -110

$\delta(^{29}\text{Si})/\text{ppm}$

Fig. 2.10 ^{29}Si NMR spectra of synthetic zeolites recorded with magic angle spinning. The resonances of Si atoms linked to n AlO$_4$ and $(4 - n)$ SiO$_4$ tetrahedra are labelled $n = 0, 1, 2, 3, 4$. The Si/Al ratios are as indicated. (Adapted from J. Klinowski, S. Ramdas, J. M. Thomas, C. A. Fyfe, and J. S. Hartman, *J. Chem. Soc., Faraday Trans. II*, 1982, **78**, 1025.)

SiO$_4$ tetrahedra, with $n = 0, 1, 2, 3, 4$. Each Al atom shifts the Si resonance by roughly +5 ppm. The relative intensities of the five peaks give the Si/Al ratio, and can be used to test model structures with different Si/Al ordering patterns.

A somewhat different use of chemical shifts is illustrated by the pH dependence of the ^{1}H NMR spectrum of the amino acid histidine. The resonance frequencies of the C2 and C4 protons in the imidazole group change smoothly with pH between the chemical shifts of the charged form HA$^+$, stable in acidic solution, and those of the neutral, deprotonated form A, which is present at high pH. At any pH, the observed chemical shift is a weighted average of the two extreme values $\delta(\text{HA}^+)$ and $\delta(\text{A})$:

$$\delta_{\text{av}} = \frac{\delta(\text{HA}^+)\,[\text{HA}^+] + \delta(\text{A})[\text{A}]}{[\text{HA}^+] + [\text{A}]}.$$

The details of this averaging process are given in Chapter 4. The midpoint of the titration occurs when the concentrations of the acid and its conjugate base are equal: $[\text{HA}^+] = [\text{A}]$, i.e. when the pH equals the pK_a of the histidine side-chain. Figure 2.11 shows the pH dependence of the chemical shift of the C2 proton of one of the histidines (His-β-146) in the oxy and deoxy forms of haemoglobin, the protein responsible for oxygen transport in blood. The pK_a of His-β-146 in the deoxy form is higher by about 1 pH unit due to the stabilization of the protonated imidazole by the CO$_2^-$ group of a nearby aspartate (Asp-β-94). These two groups are brought into close proximity by the conformational changes in the protein that accompany deoxygenation. The His-β-146–Asp-β-94 interaction is partially responsible for the change in oxygen affinity of haemoglobin with pH (the Bohr effect).

Finally, chemical shifts may be interpreted empirically using data derived from compounds of known structure. For example, Fig. 2.12 shows typical ^{1}H chemical shift ranges for assorted organic functional groups. When combined with empirical rules for predicting substituent effects (see for example Günther 1980, Bovey 1988, and Williams and Fleming 1989), such tables can be extremely useful in making connections between observed shifts and molecular structures. However, as NMR techniques

Fig. 2.11 The chemical shift of the C2 proton of histidine-β-146 in oxy- and deoxyhaemoglobin as a function of pH. (Adapted from I. D. Campbell and R. A. Dwek, *Biological spectroscopy*, Benjamin/Cummings, Menlo Park, California, 1984, p. 161.)

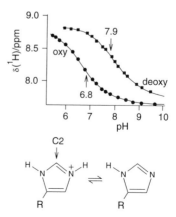

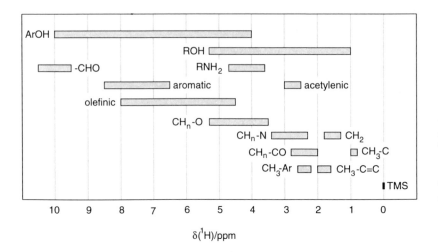

Fig. 2.12 Typical ^{1}H chemical shift ranges for some common organic functional groups. TMS = tetramethylsilane (δ = 0 ppm). More complete ^{1}H data, and tables for other nuclei, can be found in Günther (1980), Kemp (1986), and Williams and Fleming (1989).

become ever more sophisticated (Chapter 6), such methods become less and less necessary. More direct structural evidence is often provided by spin–spin couplings (Chapter 3) and nuclear Overhauser effects (Chapter 5). For example, the complete three-dimensional structures of proteins can be determined in this way (Wüthrich 1986).

Figures 2.6, 2.7, 2.8, and several others in subsequent chapters contain NMR spectra that have been drawn on a computer so as to resemble the original experimental spectra (references to which are often given in the figure captions). Genuine spectra, e.g. Figs 2.5, 2.9, 2.10, can be recognized as such by the presence of noise.

2.2 Origin of chemical shifts

A magnetic field can induce two kinds of electronic current in a molecule: *diamagnetic* and *paramagnetic* (a diamagnetic material is one in which the magnetization induced by an external field acts so as to *oppose* that field; in a paramagnetic material the induced magnetization *augments* the external field). Diamagnetic and paramagnetic currents flow in opposite directions and give rise to nuclear *shielding* and *deshielding* respectively. The shielding constant may thus be written as a sum of diamagnetic and paramagnetic contributions:

$$\sigma = \sigma_d + \sigma_p \qquad (2.7)$$

with σ_d positive, and σ_p negative.

Diamagnetic currents arise from the movement of electrons *within* atomic or molecular orbitals. As indicated in Fig. 2.2, the external field causes electrons to circulate in a plane perpendicular to the field direction. The current so induced generates a small local field opposed to B_0. The magnitude of the diamagnetic current is determined solely by the *ground state* electronic wavefunction of the atom or molecule, depends sensitively on the electron density close to the nucleus and provides the only contribution to σ for spherical, closed-shell atoms. The diamagnetic shielding is fairly easy to calculate for atoms and varies strongly with the number of electrons: 17.8 ppm for hydrogen, 261 ppm for carbon, 961 ppm for phosphorus, rising to about 10 000 ppm for atoms in the fifth row of the periodic table.

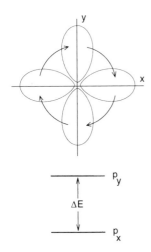

Fig. 2.13 The circulation of electronic charge brought about by the mixing of electronic wavefunctions by a magnetic field. This paramagnetic current generates a small local magnetic field that deshields the nucleus at the centre of the electron density.

Paramagnetic currents also arise from the movement of electrons within molecules, but by a more circuitous route. Imagine a somewhat artificial molecule with just two electronic states: a ground state with the form of an atomic p_x orbital containing two electrons with paired spins, and an unoccupied excited state resembling a p_y orbital (Fig. 2.13).

An external magnetic field applied along the z axis distorts the wavefunction of the ground state by mixing into it a small fraction of the excited state wavefunction. In this way, the field partially overcomes the energy gap between p_x and p_y which keeps the electrons locked along the x axis, and creates a path for electrons to circulate in the xy plane (Fig. 2.13). This induced current generates a magnetic field which (it turns out) *augments* the external field and *deshields* a nucleus at the centre of the electron density.

The extent of deshielding is clearly linked to the energy gaps involved: other things being equal, low-lying excited states should make a greater contribution than higher energy states. Theory suggests that σ_p should be approximately inversely proportional to ΔE, the average excitation energy. The paramagnetic shift is also related to the distance R between the nucleus and its surrounding electrons. Since the magnetic field at the centre of a small current loop is proportional to the inverse cube of its radius, we can expect a similar dependence for σ_p. So, very roughly,

$$\sigma_p \propto \frac{1}{\Delta E}\left\langle \frac{1}{R^3} \right\rangle \tag{2.8}$$

where $\langle \ldots \rangle$ indicates an average over the local electronic distribution.

When the electron distribution has cylindrical symmetry and the applied field is parallel to the symmetry axis, the magnetic field is unable to mix excited states into the ground state. Loosely speaking, this could be compared to a weather vane which can only rotate in the wind if it is not cylindrically symmetric about its vertical axis. Thus the π electrons in acetylene, for instance, do not give rise to paramagnetic deshielding when the magnetic field lies along the molecular axis. Similarly there is no paramagnetic contribution to the chemical shifts of atoms. The theory of diamagnetic and paramagnetic currents and the chemical shifts they generate is described by Atkins (1983).

2.3   Contributions to nuclear shielding

It should be clear from the previous section that calculating chemical shifts from first principles is a formidable task. Chemical shifts are sensitive to subtle changes in electron density, sometimes in regions of the molecule relatively remote from the nucleus of interest. One therefore needs to know the ground state wavefunction and *all* electronically excited state wavefunctions of the molecule to high accuracy—knowledge that is normally only available for the smallest of molecules. A simple illustration of the difficulties involved in computing nuclear shieldings is provided by the two iso-topomers of carbon monoxide, $^{13}C^{16}O$ and $^{13}C^{17}O$, whose ^{13}C chemical shifts differ by 0.025 ppm (a small but measurable difference). This isotope shift is due to a bond length difference of only 5×10^{-15}m. However, the situation is not nearly as gloomy as this might suggest. As we saw in Section 2.1, it is often not necessary to interpret the magnitudes of chemical shifts beyond the level of empirical correlations with structure (Fig. 2.12). Sometimes, however, chemical shift differences *can* be traced back to straightforward changes in electron density or excited state energies, as described below.

It will prove useful to divide the nuclear shielding constant σ, arbitrarily, into four parts:

$$\sigma = \text{local diamagnetic shielding}$$
$$+ \text{ local paramagnetic shielding}$$
$$+ \text{ shielding due to remote currents}$$
$$+ \text{ other sources of shielding.} \qquad (2.9)$$

The first two are contributions from the electrons in the *immediate* vicinity of the nucleus, i.e. from electrons circulating around it. The third term accounts for the diamagnetic and paramagnetic effects of electrons circulating around other (nearby) nuclei. The final part includes electric field shifts, hydrogen bonding, solvent shifts, unpaired electrons, etc., effects which, though diamagnetic or paramagnetic in origin, are more conveniently discussed separately. The remainder of this chapter gives a few illustrations of each of these contributions to σ.

Local diamagnetic shifts

The contributions to chemical shifts from local diamagnetic currents are strongly dependent on the electron density around the nucleus: the larger the electron density, the greater the shielding and the smaller the chemical shift, δ.

The ^{1}H chemical shifts of the methyl halides (Fig. 2.14) may readily be understood in these terms. As the electronegativity of the halogen increases, on going from iodine to fluorine, electron density is withdrawn from the methyl group, deshielding the protons. Indeed there is a linear correlation between the ^{1}H chemical shifts and the Pauling electronegativities of the halogens. Methane, which lacks an electron withdrawing group, has a substantially smaller chemical shift (0.13 ppm) than methyl iodide. Electropositive substituents increase the shielding still further, e.g. 0.00 ppm for $(CH_3)_4Si$.

The deshielding effect of electronegative atoms is fairly short range, as demonstrated by 1-chlorobutane, Fig. 2.15. Relative to the CH_2 group in propane, the C1 protons of 1-chlorobutane are strongly deshielded; the effect on the C2 and C3 protons is much smaller, but still measurable. The terminal methyl group (C4), four carbons away from the chlorine, is essentially unaffected by the substituent, as judged by its chemical shift relative to the methyl protons in propane.

Similar behaviour is found for monosubstituted benzenes (Fig. 2.16). Groups with electron withdrawing mesomeric effects, NO_2 and CN for

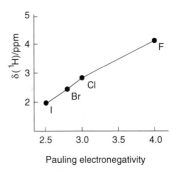

Fig. 2.14 ^{1}H chemical shifts of methyl halides plotted against the Pauling electronegativity of the halogen.

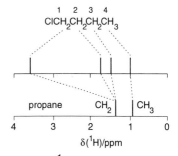

Fig. 2.15 ^{1}H chemical shifts of 1-chlorobutane compared to those of propane.

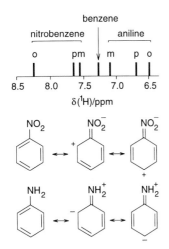

Fig. 2.16 Chemical shifts of the *ortho, meta* and *para* protons of aniline and nitrobenzene, together with resonance structures which account for the larger shielding/deshielding of the *ortho* and *para* protons compared to the *meta* proton.

example, deshield the ring protons, while electron donating groups, such as NH_2 and OCH_3, result in shielding. The shielding/deshielding is most pronounced for the protons *ortho* and *para* to the substituent, as indicated by the resonance structures.

Local paramagnetic shifts

The ΔE variation of σ_p (eqn 2.8) is nicely illustrated by the ^{59}Co chemical shifts of octahedral cobalt complexes. The five 3d orbitals of the cobalt(III) ion are split by the octahedral field of the ligands into a degenerate set of three t_{2g} orbitals and a degenerate pair of e_g orbitals, Fig. 2.17. The ground state corresponds to the $(t_{2g})^6$ electronic configuration. The four excited states arising from the first excited configuration $(t_{2g})^5(e_g)^1$ all have energies similar to Δ, the ligand field splitting. As shown in Fig. 2.17, there is a remarkably good correlation between the ^{59}Co chemical shift and the wavelength of the lowest energy absorption band (roughly proportional to $1/\Delta E$). Ligands such as carbonate, oxalate, and acetylacetonate which produce small ligand field splittings give large paramagnetic deshielding and hence large chemical shifts δ.

Fig. 2.17 ^{59}Co chemical shifts (relative to $Co(CN)_6^{3-}$) of octahedral cobalt complexes plotted against the wavelength of the first electronic absorption band. en = ethylenediamine; ox = oxalate; acac = acetylacetonate. (Adapted from R. Freeman, G. R. Murray, and R. E. Richards, *Proc. R. Soc.* (*London*), 1957, **A242**, 455.)

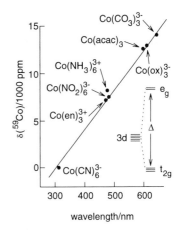

The ^{13}C chemical shifts of monosubstituted benzenes provide a good example of the dependence of σ_p on $\langle R^{-3} \rangle$. As shown in Fig. 2.18, the *para* carbon is deshielded by electron withdrawing substituents and shielded by electron releasing groups. Although this is the same trend observed for the ^{1}H shifts in these compounds, the effect is paramagnetic rather than diamagnetic in origin. Electron donating groups delocalize their lone pairs into the ring and increase the electron density at the *ortho* and *para* carbons. The increased electron repulsion causes the orbitals around these atoms to *expand*, reducing $\langle R^{-3} \rangle$ and hence δ.

At this point we can see why the chemical shift range of protons is so small compared to other nuclei ($\approx$10 ppm for ^{1}H; $\approx$200 ppm for ^{13}C; $\approx$300 ppm for ^{19}F, $\approx$500 ppm for ^{31}P, etc.). Local diamagnetic and local paramagnetic currents make only modest contributions to ^{1}H shielding because of the low electron densities and high electronic excitation energies associated with hydrogen atoms. Indeed, ^1H chemical shifts are often more strongly influenced by the diamagnetic and paramagnetic currents in *neighbouring groups* of atoms which have larger electron densities and lower excitation energies, as we shall now see.

Neighbouring groups

To see how a nucleus can be shielded or deshielded by the motions of electrons on nearby groups of atoms, consider the following highly simplified model. Suppose that the distribution of electrons in the neighbouring group has cylindrical symmetry (e.g. the triple bond in acetylene) and pretend that the magnetic field generated by the circulating electrons has the same form as the field that would be produced by a magnetic dipole (i.e. a tiny bar magnet) sitting at the centre of the electron density. (Never mind that this approximation is only strictly valid at distances much larger than the extent of the electron distribution.) Let the size of the induced magnetic moment be μ_{par} when B_0 is parallel to the symmetry axis, and μ_{per}, when it is perpendicular: only when the neighbouring group has very high symmetry, e.g. tetrahedral, will the two magnetic moments be equal.

The properties of magnetic dipoles are explored in Section 3.8. For now, we simply need to recognize that the field generated by a magnetic dipole μ has the form shown in Fig. 3.32(a) (p. 40), and that its component along the B_0 direction at a point specified by polar coordinates (r,θ) (see Fig. 3.31, p. 40) is

$$\left(\frac{\mu_0}{4\pi}\right)\left(\frac{\mu}{r^3}\right)(3\cos^2\theta-1) \tag{2.10}$$

(μ_0 is the vacuum permeability). Note that when $\theta = 0°$, the dipolar field is $2(\mu_0/4\pi)(\mu/r^3)$, and when $\theta = 90°$, it is $-(\mu_0/4\pi)(\mu/r^3)$, i.e. half as big and in the opposite direction.

To discover whether the field arising from the induced magnetic moment shields or deshields a nearby nucleus, we need simply to average the dipolar field experienced by that nucleus over all orientations of the molecule as it tumbles in solution. This is done in Fig. 2.19 together with Table 2.1.

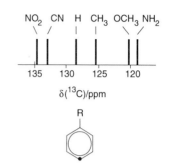

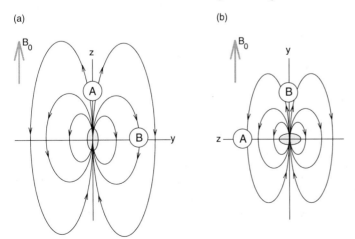

(a)　　　　　　　(b)

Fig. 2.18 ^{13}C chemical shifts of the *para* carbons in monosubstituted benzenes.

Fig. 2.19 Dipolar lines of flux generated by an induced magnetic moment at the centre of a cylindrically symmetric neighbouring group (shown as an elipsoid) with two nearby protons A and B. The external field B_0 is parallel in (a) and perpendicular in (b) to the symmetry axis of the group. The diagram has been drawn for the case $\mu_{par} > \mu_{per} > 0$. See also Table 2.1.

Table 2.1 Nuclear shielding/deshielding arising from neighbouring group anisotropy.

B_0 parallel to	θ_A	θ_B	Dipolar field at H_A	Dipolar field at H_B
x	90°	90°	$-\mu_{per}$	$-\mu_{per}$
y	90°	0	$-\mu_{per}$	$2\mu_{per}$
z	0	90°	$2\mu_{par}$	$-\mu_{par}$
Total dipolar field			$2(\mu_{par} - \mu_{per})$	$-(\mu_{par} - \mu_{per})$

To obtain the net dipolar field at the positions of two protons H_A and H_B in a molecule tumbling end over end in solution (see Fig. 2.19), add up the fields experienced by each spin when the external field B_0 is parallel to the x, y and z axes of the neighbouring group. θ_A and θ_B are the angles between B_0 and vectors connecting the centre of the neighbouring group to the two protons. The dipolar fields are quoted in units of $(\mu_0/4\pi)r^{-3}$ (see eqn 2.10). μ_{par} and μ_{per} are the magnetic moments induced in the neighbouring group by the external field when it is respectively parallel and perpendicular to the symmetry axis of the group.

A proton H_A, lying on the symmetry axis of the neighbouring group, experiences an average field $2(\mu_0/4\pi)(\mu_{par} - \mu_{per})/r^3$; the corresponding field for a proton H_B, situated to one side of the group, is $-(\mu_0/4\pi)(\mu_{par} - \mu_{per})/r^3$. That is, the size of the local field depends on the magnetic *anisotropy* of the neighbouring group ($\mu_{par} - \mu_{per}$) and has opposite signs for the two nuclei. This source of nuclear shielding/deshielding is therefore often referred to as *neighbouring group anisotropy*. When $\mu_{par} > \mu_{per}$, the average dipolar field at H_A augments B_0, while that at H_B opposes B_0: H_A is therefore deshielded, and H_B shielded by the electrons circulating around the neighbouring group. The situation is reversed if $\mu_{par} < \mu_{per}$. Note that *diamagnetic* currents in the neighbouring group correspond to a *negative* induced magnetic moment (opposing B_0), while *paramagnetic* currents give *positive* μ_{par} and μ_{per}.

A simple extension of this argument leads to the sketches in Fig. 2.20. When $\mu_{par} > \mu_{per}$ (a), a nucleus lying *within* one of the two cones (half angle 54.7°) should be deshielded, while a nucleus in the region *outside* the cones should be shielded. The opposite holds when $\mu_{par} < \mu_{per}$ (b). Of course, both the angular and radial dependence of the induced field are more complicated than the dipolar approximation would suggest; nevertheless this model remains a useful way of thinking about neighbouring group anisotropy, and if approached with caution, can be used to predict the direction, if not the magnitude, of chemical shifts.

As an example of neighbouring group anisotropy, consider the triple bond in acetylene. The dominant contribution to the shielding of nearby nuclei comes from the *paramagnetic* currents that are induced when the field is *perpendicular* to the axis of the molecule: this is because the diamagnetic currents are almost isotropic, and the paramagnetic current around the molecular axis is zero due to the cylindrical symmetry. Thus $\mu_{per} > 0$, and

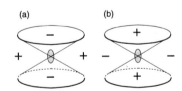

(a)   (b)

Fig. 2.20 Regions of shielding (+) and deshielding (−) due to neighbouring group magnetic anisotropy, for (a) $\mu_{par} > \mu_{per}$ and (b) $\mu_{par} < \mu_{per}$. The half angle of the cones is 54.7° (the so-called *magic angle* at which $3\cos^2\theta = 1$). The axis of the cones coincides with the symmetry axis of the neighbouring group (shown as an ellipsoid).

$\mu_{par} - \mu_{per} < 0$ so that, from Fig. 2.20(b), the protons in acetylene should be *shielded*.

Figure 2.21 shows the ^{1}H chemical shifts of acetylene, ethylene, and ethane. Based on the hybridization of the molecular orbitals, one would expect $\delta(C_2H_2) > \delta(C_2H_4) > \delta(C_2H_6)$: as the s-electron character increases in the order $sp^3 < sp^2 < sp$, so the bonding electrons should be held more tightly to the carbon, removing electron density from the hydrogen atoms and deshielding the protons. The observed order of shifts arises from the *shielding* effect of the π electrons of the triple bond, as discussed above, and also the neighbouring group effect of the C=C double bond which *deshields* the protons in ethylene.

A further example of the neighbouring group effect of a C≡C bond is provided by the acetylene-substituted phenanthrene in Fig. 2.22. The proton beside the triple bond falls in the deshielding region of Fig. 2.20(b) and suffers a shift of +1.7 ppm relative to the corresponding proton in phenanthrene itself.

A more important instance of the neighbouring group effect occurs in *aromatic* compounds, whose extensive π electron 'clouds' can support large electronic currents. In molecules such as benzene, the dominant contribution to the magnetic anisotropy comes from the circulation of the π electrons *within* their delocalized molecular orbitals; i.e. the neighbouring group effect is due principally to the *diamagnetic* moment induced when the external field is perpendicular to the molecular plane, as indicated in Fig. 2.23. However, the end result is the same as for acetylene because the induced diamagnetic moment is *opposed* to the external field ($\mu_{par} < 0$), so that the anisotropy $\mu_{par} - \mu_{per}$ is once again negative. Thus we can anticipate deshielding for nuclei in the plane of the aromatic ring, and shielding for any nuclei above or below the ring.

This *ring current shift* is demonstrated clearly by benzene itself, whose ^{1}H chemical shift is 1.4 ppm to the high frequency side of the olefinic protons in cyclohexa-1,3-diene, Fig. 2.24(a). A more interesting example is the dimethyl-substituted pyrene which, with 14 π electrons, is aromatic according to the $4n + 2$ rule ($n = 3$). The ring protons are deshielded, as expected, by analogy with benzene, but the methyl groups, which protrude above and below the plane of the molecule, are shielded by more than 5 ppm relative to ethane, Fig. 2.24(b). (Negative δ values correspond to protons resonating to the low frequency side of TMS.) As indicated in Fig. 2.23, these protons lie in a region where the induced field *opposes* the external field.

Finally, the planar annulene in Fig. 2.24(c), with $4n + 2 = 18$ π electrons ($n = 4$), shows two proton resonances, one from the 12 strongly deshielded external protons, and one at −2.99 ppm from the six internal protons. The latter set of hydrogen atoms lies within the current loop formed by the circulating π electrons, and as indicated in Fig. 2.23 is shielded by the ring current effect. Note that the chemical shifts of these nuclei cannot, even qualitatively, be described using the point dipole approximation (Fig. 2.20)

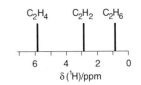

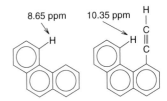

Fig. 2.21 ^{1}H chemical shifts of ethane, ethylene, and acetylene.

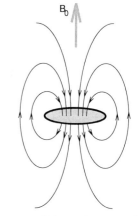

Fig. 2.22 Deshielding due to the magnetic anisotropy of a C≡C bond.

Fig. 2.23 Schematic magnetic flux lines arising from the diamagnetic current induced in an aromatic ring when the external field is perpendicular to the plane of the ring.

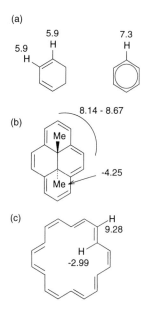

Fig. 2.24 Ring current shifts. ^{1}H chemical shifts in:
(a) benzene compared to cyclohexa-1,3-diene; (b) a dimethyl-substituted pyrene; (c) [18]-annulene.

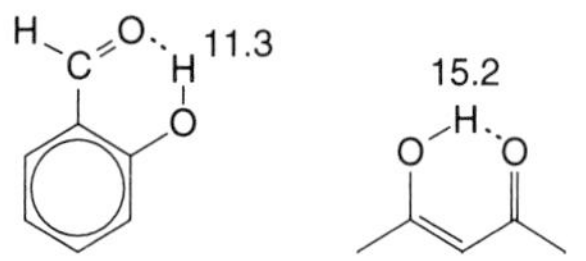

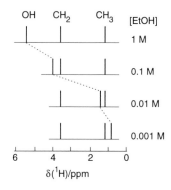

Fig. 2.25 Deshielding due to hydrogen bonding. ^{1}H chemical shifts in salicylaldehyde and the enol form of acetylacetone.

which predicts deshielding for any nucleus in the plane of the aromatic ring, however close to the centre of the molecule.

The magnitude of the neighbouring group effect depends on the magnetic anisotropy of the group itself and not on the nucleus being shielded or deshielded. Neighbouring group anisotropy is therefore *relatively* more important for protons with their small local diamagnetic and paramagnetic (de)shieldings than for other nuclei, ^{13}C for instance, which have larger electron densities and lower excitation energies.

Although only triple bonds and aromatic rings have been discussed, neighbouring group effects exist for other functional groups—C–C, C=C, C=O, N=O, for example—and have a strong influence on ^{1}H chemical shifts.

Other sources of chemical shifts

Hydrogen bonding is responsible for some of the largest observed deshielding effects in ^{1}H NMR. Two compounds that form intramolecular hydrogen bonds are shown in Fig. 2.25: in both, the hydrogen-bonded proton is heavily deshielded. Intermolecular hydrogen bonds, which are generally somewhat weaker, produce smaller shifts. For example, the hydroxyl proton resonance of ethanol moves by about −4 ppm when the intermolecular hydrogen bonding is disrupted by dilution into CCl$_4$ (see Fig. 2.26). The limiting shift of the OH resonance at infinite dilution (0.8 ppm) compares well with the value for monomeric ethanol in the gas phase (0.55 ppm). Similar changes in the OH chemical shift are found with increasing temperature, which also favours the monomer side of the monomer ⇌ dimer, trimer, ... equilibrium. The origin of the deshielding caused by hydrogen bonding is unclear. Most probably, the strong electric field of the Y atom in X–H···Y draws the hydrogen atom away from the electrons in the X–H bond, so reducing the electron density immediately around it.

Chemical shifts are also affected by the local *electric* fields arising from charged or polar groups. These can modify both diamagnetic and paramagnetic currents by polarizing local electron distributions, and by perturbing ground and excited state wavefunctions and energies. Positive charges usually deshield nearby protons, while negative charges often give rise to shielding. For example, the protons on the imidazole side-chain of the

Fig. 2.26 Hydrogen bonding shifts in ethanol, as a function of concentration in CCl$_4$.

amino acid histidine shift to higher frequencies by about 1 ppm when the ring is protonated (see Fig. 2.11).

Finally, in this far from complete survey, there are the *paramagnetic* shifts produced by unpaired electrons (in this context paramagnetic refers to the permanent magnetic moment of the electron and not to the induced electronic currents discussed earlier). Unpaired electrons give rise to large dipolar magnetic fields—the gyromagnetic ratio of the electron is 660 times that of the proton—which can result in substantial nuclear shielding/deshielding, as illustrated by the two metal–porphin complexes in Fig. 2.27. The two types of proton in the aromatic ligand, which have large ring current shifts in the diamagnetic Zn^{2+} complex, are heavily shielded in the paramagnetic Fe^{3+} compound.

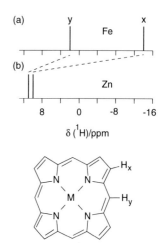

Fig. 2.27 ^1H chemical shifts of the two types of proton (x and y) in (a) the paramagnetic Fe^{3+} (porphin) (CN^-) and (b) the diamagnetic Zn^{2+} (porphin).

3 Spin–spin coupling

Chapter 2 may have given the impression that the appearance of NMR spectra is determined solely by chemical shifts—one resonance for each distinct nuclear environment. In fact, there is another extremely valuable source of information encoded in most NMR spectra, namely the magnetic interactions between nuclei, known variously as *spin–spin couplings, scalar couplings,* or *J-couplings.* Amongst other things, these interactions cause the ^1H spectrum of liquid ethanol to comprise not three but eight (and sometimes more) resonances when recorded at high resolution (Fig. 3.1).

Fig. 3.1 400 MHz ^1H NMR spectrum of liquid ethanol showing the splittings produced by spin–spin coupling. Compare this spectrum with Fig. 1.2, in which the fine structure is obscured by instrumental linebroadening.

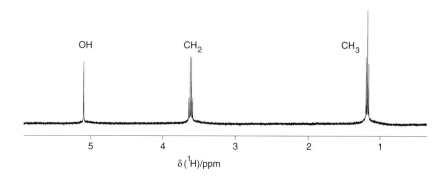

3.1 Effect on NMR spectra

As Fig. 3.1 suggests, nuclear spin–spin coupling causes NMR lines to split into a small number of components with characteristic relative intensities and spacings. In the case of ethanol, the CH_3 peak becomes a *triplet*—three equally spaced lines with relative amplitudes in the ratio 1 : 2 : 1—while the CH_2 resonance is split into a *quartet*—four equally spaced lines with relative intensities 1 : 3 : 3 : 1. To see how this *multiplet* structure arises we focus initially on a much simpler molecule, the formate ion HCO_2^- in which the carbon is ^{13}C.

There are just two magnetic nuclei in $H^{13}CO_2^-$ (assuming the oxygens are ^{16}O): the ^1H and the ^{13}C, both of which have spin quantum number $\frac{1}{2}$. On the basis of the previous chapter, we might expect to see a single line in both the proton and carbon NMR spectra. In fact both spectra contain *two* lines (a *doublet*), as shown in Fig. 3.2. The doublet splitting (195 Hz in this case) gives the strength of the ^1H–^{13}C spin–spin interaction and is the same in the two spectra.

The ^1H resonance is split into two because the magnetic moment of the ^{13}C nucleus is the source of a small local magnetic field whose direction is

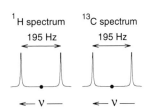

Fig. 3.2 ^1H and ^{13}C NMR spectra of $H^{13}CO_2^-$ showing the splittings produced by ^1H–^{13}C scalar coupling. The two chemical shift positions are indicated by dots.

determined by the ^{13}C *magnetic quantum number*. When the ^{13}C is in its $m = +\frac{1}{2}$ state (here denoted C↑) its magnetic field at the position of the ^{1}H opposes the external field and shifts the ^{1}H resonance to lower frequency, as indicated in Fig. 3.3. Conversely, for an $m = -\frac{1}{2}$ carbon (C↓), the local field adds to the external field at the proton and moves its resonance to higher frequency. In the language of chemical shifts, a C↑ carbon shields the proton and a C↓ deshields it. The two components of the ^{1}H doublet thus correspond to two sorts of H^{13}CO$_2^-$ molecule: those with C↑ and those with C↓. Since the difference in energy between the two configurations of the carbon nucleus is tiny compared to the thermal energy kT, the two kinds of H^{13}CO$_2^-$ are equally likely, and the two components of the ^{1}H doublet are equally intense. An exactly analogous argument explains the splitting of the ^{13}C resonance by the ^{1}H.

It is evident from Fig. 3.3 that the spin–spin interaction in H^{13}CO$_2^-$ stabilizes the antiparallel arrangements of nuclear spins (H↑C↓ and H↓C↑) and destabilizes the parallel configurations (H↑C↑ and H↓C↓).Thus the two energy levels of the proton are each split into two, with energies determined by the relative orientations of the ^{13}C and ^{1}H spin angular momenta. Molecules containing a C↑ undergo ^{1}H NMR transitions (H↑C↑ → H↓C↑) at a frequency *lower* than the chemical shift, because the transition is from an energetically unfavourable state (parallel spins) to a favourable one (antiparallel spins). Conversely, molecules containing C↓ resonate to *higher* frequency (H↑C↓ → H↓C↓).

A *heteronuclear* example has been used to illustrate the nature of spin–spin coupling merely as a matter of convenience; *homonuclear* couplings, e.g. between protons with different chemical shifts, give rise to splittings in exactly the same way.

The properties of spin–spin coupling as revealed by H^{13}CO$_2^-$ may be summarized and generalized in the following simple expression for the energy of two interacting nuclei A and X (not necessarily spin-$\frac{1}{2}$):

$$E = h\, J_{AX}\, m_A\, m_X \tag{3.1}$$

in which m_A and m_X are the magnetic quantum numbers of the two nuclei and J_{AX} is the *spin–spin coupling constant*. J_{AX} is measured in frequency units (Hertz) and may be positive or negative: if the antiparallel arrangement of nuclear spins is energetically favoured, then $J_{AX} > 0$ (as in H^{13}CO$_2^-$); when the parallel spin configuration is lower in energy, $J_{AX} < 0$.

Combining eqn 3.1 with the selection rule $\Delta m_A = +1$, one can see that the AX interaction shifts the NMR frequency of spin A by an amount $-J_{AX}m_X$. Together with eqn 2.3, this leads straightforwardly to a general resonance condition for spin A:

$$\nu_A = \frac{\gamma_A B_0(1 - \sigma_A)}{2\pi} - \sum_{X \ne A} J_{AX}m_X \tag{3.2}$$

where the summation runs over all spins (X) with appreciable spin–spin coupling to A. Figure 3.4 shows the complete energy level diagram, together with the corresponding NMR spectra, for a pair of spin-$\frac{1}{2}$ nuclei with (c) and

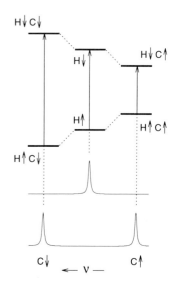

Fig. 3.3 The effect of ^{1}H–^{13}C scalar coupling in H^{13}CO$_2^-$ on the energy levels and spectrum of the ^{1}H nucleus. For clarity, the energy-level shifts due to the spin–spin coupling have been greatly exaggerated. The central pair of energy levels and the upper spectrum are appropriate in the *absence* of a spin–spin interaction. Scalar coupling produces the energy levels on the left and right, and the lower spectrum.

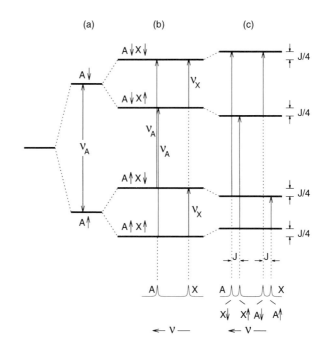

Fig. 3.4 Energy levels and spectra of a pair of spins, A and X. From left to right, magnetic interactions are introduced in the order: (a) the interaction of A with the magnetic field B_0; (b) the interaction of X with B_0; (c) the scalar coupling $J = J_{AX} > 0$. Energies are given in frequency units. The energy-level shifts due to the spin–spin coupling have been greatly exaggerated for clarity. ν_A and ν_X are the resonance frequencies of the two spins.

without (b) scalar coupling. Note that the allowed transitions are those in which a *single* spin flips: the transitions A↑X↑ ↔ A↓X↓ and A↑X↓ ↔ A↓X↑ are forbidden.

It should be clear from eqns 3.1 and 3.2 that the *sign* of the coupling constant has no effect on the appearance of the spectrum. For example, changing J_{AX} in Fig. 3.4 from positive to negative simply interchanges the two components of each doublet: the A resonance of the X↑ molecule swaps places with that of the X↓ molecule, and similarly in the X spectrum.

These simple ideas, exemplified by $H^{13}CO_2^-$, and embodied in eqns 3.1 and 3.2, allow one to predict the effect of spin–spin coupling on the NMR spectrum of almost any molecule. The exceptions will be dealt with later (principally Sections 3.4 and 3.5).

3.2 Multiplet patterns

Having seen that coupling between nuclear spins can affect NMR spectra, we now look at some frequently encountered spin systems (collections of coupled nuclei) to see how they give rise to distinctive multiplets (doublets, triplets, quartets, etc.).

At this stage it is assumed that all pairs of spins are *weakly coupled*, i.e. that the difference in resonance frequencies of the two nuclei greatly exceeds their mutual coupling (the complications associated with *strong coupling* are discussed in Section 3.5). All nuclei are spin-$\frac{1}{2}$, unless otherwise stated. The term *equivalent nuclei* is used to describe spins in identical environments, with identical chemical shifts—for example the protons in CH_4 or the fluorines in CF_3COOH. This somewhat loose definition of equivalence will be refined at the end of this section.

The following paragraphs deal with the effect of nuclei M and X on the NMR signal of nucleus A. According to convention, spins with very different chemical shifts are labelled by letters far apart in the alphabet (e.g. A, M, X). Nuclei having similar shifts, and thus likely to be strongly coupled, are assigned adjacent letters in the alphabet (e.g. A, B, C).

Finally it must be said that the predictions discussed below are not absolutely infallible: the expected multiplet patterns may be obscured if the splitting is smaller than the linewidth (see Chapter 5), or modified if the molecule is undergoing a dynamic process capable of averaging spin–spin interactions (see Chapter 4).

Coupling to a single spin-$\frac{1}{2}$ nucleus (AX)

As already discussed for $H^{13}CO_2^-$, the interaction of nucleus A with a single spin-$\frac{1}{2}$ nucleus, X, causes the A resonance to split into two equally intense lines centred at the chemical shift of A (a doublet), with spacing equal to the AX coupling constant, J_{AX}. The interaction is symmetrical, so that the spectrum of X is also a doublet, with the same splitting (Figs. 3.2–3.4).

Coupling to two inequivalent spin-$\frac{1}{2}$ nuclei (AMX)

A step up from the previous case is the AMX spin system, which consists of three nuclei with different chemical shifts and three distinct coupling constants: J_{AM}, J_{AX}, J_{MX}. Equation 3.2 can be used to predict the spectrum of A, by drawing up a list of the possible values of the magnetic quantum numbers of M and X (Table 3.1). Four lines are expected because there are four non-degenerate arrangements of the M and X spins (M↑X↑, M↑X↓, M↓X↑, M↓X↓). These peaks are displaced from the chemical shift of A by simple combinations of the couplings to spin A (J_{AM} and J_{AX}). The A multiplet should therefore be a *doublet of doublets*, as shown in Fig. 3.5.

A different way to see how this pattern arises is to construct the spectrum in stages. Imagine first of all that both J_{AM} and J_{AX} are zero, so that the A spectrum is a singlet at the chemical shift position. Now suppose the AM

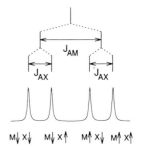

Fig. 3.5 NMR spectrum of nucleus A in an AMX spin system. The four components of the A multiplet, a doublet of doublets, arise from the four combinations of M and X spins, indicated ↑ ($m = +\frac{1}{2}$) and ↓ ($m = -\frac{1}{2}$). Drawn for $J_{AM} > J_{AX} > 0$.

Table 3.1 Spin–spin coupling in an AMX spin system

m_M	m_X	$-\sum\limits_{K=M,X} J_{AK}m_K$
$+\frac{1}{2}$	$+\frac{1}{2}$	$-\frac{1}{2}(J_{AM} + J_{AX})$
$+\frac{1}{2}$	$-\frac{1}{2}$	$-\frac{1}{2}(J_{AM} - J_{AX})$
$-\frac{1}{2}$	$+\frac{1}{2}$	$+\frac{1}{2}(J_{AM} - J_{AX})$
$-\frac{1}{2}$	$-\frac{1}{2}$	$+\frac{1}{2}(J_{AM} + J_{AX})$

The final column shows the shift in the resonance frequency of A for each of the four spin configurations of M and X, both of which have $I = \frac{1}{2}$ (see eqn 3.2).

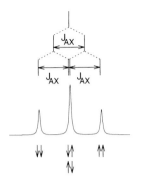

Fig. 3.6 NMR spectrum of nucleus A in an AX_2 spin system. The triplet arises from the four combinations of the two X spins, as indicated. Drawn for $J_{AX} > 0$.

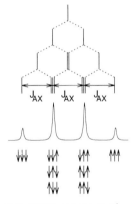

Fig. 3.7 NMR spectrum of nucleus A in an AX_3 spin system. The quartet arises from the eight combinations of the three X spins, as indicated. Drawn for $J_{AX} > 0$.

coupling is 'switched on', to give a doublet with splitting J_{AM}. Finally, when the AX coupling is introduced, each of the lines of the doublet is itself split into a doublet, with splitting J_{AX}. This stepwise procedure, probably the quickest way of arriving at multiplet patterns, is illustrated in Fig. 3.5. The order in which the couplings are introduced is irrelevant. Of course, the exact appearance of the doublet of doublets will depend on the values of the coupling constants. This point is illustrated later (Fig. 3.16) for a *four* spin system.

Coupling to two equivalent spin-$\frac{1}{2}$ nuclei (AX_2)

This is a special case of the AMX spin system, with $J_{AM} = J_{AX}$. As may be seen from Fig. 3.6 (and Table 3.1), the two central lines of the doublet of doublets coincide to give a *triplet* centred at the chemical shift of A, with linespacing equal to the coupling constant, and relative intensities $1:2:1$. The central line of the triplet thus arises from two degenerate arrangements of the X spins ($\uparrow\downarrow$ and $\downarrow\uparrow$), in both of which the local magnetic fields due to the two X nuclei exactly cancel.

Coupling to three equivalent spin-$\frac{1}{2}$ nuclei (AX_3)

The multiplet pattern of A in an AX_3 spin system (three identical AX coupling constants) is a four line quartet (Fig. 3.7 and Table 3.2): two peaks displaced from the chemical shift position by $\pm\frac{3}{2}J_{AX}$, and two peaks with three times the intensity, at $\pm\frac{1}{2}J_{AX}$. The line at $-\frac{1}{2}J_{AX}$, for example, has relative intensity 3 because there are three ways of finding two X nuclei with spin $\uparrow$ and one with spin $\downarrow$.

Table 3.2 Spin–spin coupling in an AX_3 spin system.

m_1	m_2	m_3	$-\sum_i J_{AX}m_i$
$+\frac{1}{2}$	$+\frac{1}{2}$	$+\frac{1}{2}$	$-\frac{3}{2}J_{AX}$
$+\frac{1}{2}$	$+\frac{1}{2}$	$-\frac{1}{2}$	
$+\frac{1}{2}$	$-\frac{1}{2}$	$+\frac{1}{2}$	$-\frac{1}{2}J_{AX}$
$-\frac{1}{2}$	$+\frac{1}{2}$	$+\frac{1}{2}$	
$+\frac{1}{2}$	$-\frac{1}{2}$	$-\frac{1}{2}$	
$-\frac{1}{2}$	$+\frac{1}{2}$	$-\frac{1}{2}$	$+\frac{1}{2}J_{AX}$
$-\frac{1}{2}$	$-\frac{1}{2}$	$+\frac{1}{2}$	
$-\frac{1}{2}$	$-\frac{1}{2}$	$-\frac{1}{2}$	$+\frac{3}{2}J_{AX}$

The final column shows the shift in the resonance frequency of A for each of the eight spin configurations of the three X spins ($I = \frac{1}{2}$), labelled 1, 2 and 3 (see eqn 3.2).

Coupling to n equivalent spin-$\frac{1}{2}$ nuclei (AX$_n$)

It should be clear how the results for AX, AX$_2$ and AX$_3$ can be generalized. For n equivalent X (spin-$\frac{1}{2}$) nuclei, the A resonance is split into $n + 1$ equally spaced lines, with relative intensities given by simple combinatorial arithmetic. The amplitude of the mth line ($m = 0,1,2, ... n$) of an AX$_n$ multiplet is simply the number of ways of finding m spins $\uparrow$ and the remainder $\downarrow$, i.e. $n!/m!(n–m)!$. To put it another way, the amplitudes are given by the coefficients in the binomial expansion of $(1+x)^n$, or, equivalently by the $(n+1)$th row of Pascal's triangle (Fig. 3.8).

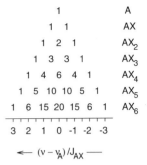

			1					A
		1		1				AX
		1	2		1			AX$_2$
	1		3	3		1		AX$_3$
	1	4	6		4	1		AX$_4$
1		5	10	10	5		1	AX$_5$
1	6	15	20	15	6	1		AX$_6$

3   2   1   0   -1   -2   -3

$\longleftarrow (\nu - \nu_A)/J_{AX} \longrightarrow$

Fig. 3.8 Pascal's triangle showing the binomial coefficients in the expansion of $(1 + x)^n$. The rows give the relative intensities of the $(n + 1)$ lines in the A multiplet of an AX$_n$ spin system ($n = 0$–6), where X is a spin-$\frac{1}{2}$ nucleus. The columns give the positions of the lines, relative to the chemical shift position, in units of J_{AX}.

Coupling involving $I > \frac{1}{2}$ nuclei

If the nucleus of interest, A, has spin quantum number greater than $\frac{1}{2}$, its multiplet structure can be predicted in exactly the same way as for a spin-$\frac{1}{2}$ nucleus. This should be evident from eqns 3.1 and 3.2, and is demonstrated, for the case of a spin-1 coupled to a spin-$\frac{1}{2}$, in Fig. 3.9. For example, the ^{14}N ($I = 1$) and ^{15}N ($I = \frac{1}{2}$) NMR spectra of, respectively, ^{14}NH$_4^+$ and ^{15}NH$_4^+$ both consist of a quintet, with relative peak intensities $1 : 4 : 6 : 4 : 1$. The NH coupling constants of the two isotopomers are in the ratio $0.713 : 1$, which is the ratio of the gyromagnetic ratios of the two nitrogen isotopes (see Table 1.3).

However, nuclei with $I > \frac{1}{2}$ possess, in addition to their magnetic dipole moment, an *electric quadrupole moment* that can interact with local *electric field gradients*. For molecules tumbling in solution, this interaction can lead to efficient relaxation of the quadrupolar nucleus and NMR lines that are so broad that the expected multiplet patterns are partially or completely obscured. The quadrupolar relaxation mechanism is discussed further in Section 5.6.

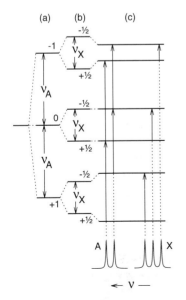

Fig. 3.9 Energy levels and spectra of a spin-1 nucleus (A) coupled to a spin-$\frac{1}{2}$ (X). From left to right, magnetic interactions are introduced in the order: (a) the interaction of A with the magnetic field B_0; (b) the interaction of X with B_0; (c) the scalar coupling ($J = J_{AX} < 0$). Note that the spectrum of A is a doublet because its four allowed NMR transitions are pairwise degenerate. The X spectrum comprises three lines arising from the $m = +1, 0, -1$ states of A. For clarity, the energy-level shifts due to the spin–spin coupling ($hJ_{AX}m_Am_X$) have been greatly exaggerated. ν_A and ν_X are the resonance frequencies of the two spins. Energy levels are labelled with the appropriate magnetic quantum numbers.

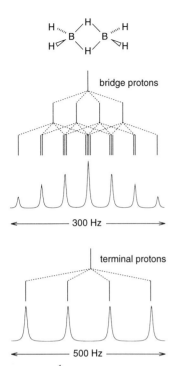

bridge protons

← 300 Hz →

terminal protons

← 500 Hz →

Fig. 3.10 ^{1}H NMR spectra of the terminal and bridge protons in diborane, $^{11}B_2H_6$.

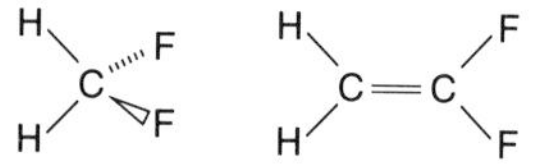

Fig. 3.11 CH_2F_2 (magnetically equivalent protons) and $CH_2 = CF_2$ (chemically equivalent protons).

For A ($I = \frac{1}{2}$) coupled to X ($I > \frac{1}{2}$), the principles established above for spin-$\frac{1}{2}$ nuclei can easily be extended. The magnetic moment of a spin-I particle has $2I + 1$ orientations with respect to the magnetic field B_0 (eqn 1.4): therefore, a nucleus coupled to a single X spin with quantum number I should show a multiplet comprising $2I + 1$ lines with equal spacings and amplitudes. For example, the ^{13}C spectrum of deuterated chloroform, ^{13}CDCl$_3$, is a 1 : 1 : 1 triplet arising from the three equally probable states of the deuteron, $m = +1, 0, -1$ (Fig. 3.9). Once again, quadrupolar relaxation may upset these predictions. Rapid relaxation of the quadrupolar nucleus may have the effect of 'decoupling' A and X, such that no splitting is observed in the spectrum of A. For example, ^{35}Cl and ^{37}Cl (both $I = \frac{3}{2}$) rarely produce splittings in the NMR spectra of nearby nuclei. We shall return to this point in Section 5.6.

For coupling to *equivalent* $I > \frac{1}{2}$ nuclei, the multiplet patterns are easily deduced using the 'tree diagram' approach introduced in Fig. 3.5. For instance, the terminal protons of ^{11}B$_2$H$_6$ (diborane) show a 1 : 1 : 1 : 1 quartet due to coupling to the directly bonded ^{11}B ($I = \frac{3}{2}$), while the bridge protons exhibit a seven line pattern with relative intensities 1 : 2 : 3 : 4 : 3 : 2 : 1, arising from equal interactions with the two symmetrically placed borons (Fig. 3.10).

Equivalent nuclei

Up to now, we have used the term *equivalent* somewhat loosely to describe nuclei with identical chemical shifts, usually as a result of molecular symmetry. In fact there are two kinds of equivalence: chemical and magnetic. The distinction is best seen by means of an example. Consider the protons in the two compounds CH_2F_2 and $CH_2=CF_2$ (Fig. 3.11) and their couplings to the two fluorines F_a and F_b. In CH_2F_2, the two protons have the same chemical shift *and* each has identical couplings to F_a and to F_b: as such they are termed *magnetically equivalent*. The same cannot be said of $CH_2=CF_2$, where the *cis* and *trans* ^{1}H–^{19}F coupling constants differ: in this case the protons are said to be *chemically equivalent*.

More generally, a set of nuclei with identical chemical shifts (call them a, b, c, ...) are magnetically equivalent if, for every other nucleus (e.g. z) in the molecule, the spin–spin coupling constants satisfy a relation:

$$J_{az} = J_{bz} = J_{cz} = \dots . \tag{3.3}$$

As might be expected, the NMR spectra of molecules containing chemically equivalent spins are rather more complex than for similar compounds with magnetically equivalent nuclei. For example, the ^{1}H spectrum of $CH_2=CF_2$ has no fewer than ten lines. The analysis of such spectra is not straightforward and will not be attempted here: a good discussion is given by Günther (1980). Fortunately, chemical equivalence is rather less common than magnetic equivalence, to which we now turn.

The ^{1}H spectrum of CH_2F_2 comprises just three lines: a 1 : 2 : 1 triplet with splitting equal to the proton–fluorine coupling constant J_{HF} (the spin quantum number of ^{19}F is $\frac{1}{2}$). The remarkable thing about this spectrum is not the triplet, which is exactly what one would expect for a *single* proton coupled to two identical fluorines, but the *absence* of any splittings arising from the proton–proton coupling. Although the two protons interact with one another, this coupling does not manifest itself as a splitting in the spectrum. In fact this is a general feature of scalar coupling: *spin–spin interactions within a group of magnetically equivalent nuclei do not produce multiplet splittings.*

Perhaps without realizing it, we have already seen several instances of this phenomenon: each of the five molecules in Fig. 2.5 contains a single group of (magnetically) equivalent protons and each gives rise to an NMR singlet. A more esoteric example is the highly symmetrical molecule dodecahedrane (Fig. 3.12) whose ^1H spectrum also consists of a single peak.

The high resolution spectrum of ethanol in Fig. 3.1 can now be understood. The ethyl protons make up an A_3X_2 spin system: the triplet arises because each of the CH_3 protons couples equally to the two equivalent CH_2 protons, while the quartet comes from the CH_2 protons interacting identically with each of the CH_3 protons. As discussed in Chapter 4, the rapid internal rotation around the C–C bond averages out the chemical shift differences associated with the different conformations of the molecule, and effectively renders the three methyl protons magnetically equivalent, and similarly the two methylene protons. The absence of splittings from coupling between the CH_2 group and the OH proton is another story, also told in Chapter 4.

In Section 3.4, we shall see *why* magnetically equivalent nuclei do not split one another's NMR lines, but first a few examples that illustrate how multiplet patterns can be used to determine or verify the structures of molecules, without prior knowledge of the magnitudes of the chemical shifts or coupling constants involved. For convenience the multiplet patterns expected for some of the simple spin systems discussed above are collected together in Fig. 3.13.

$\delta(^1H) = 3.38$ ppm

Fig. 3.12 Dodecahedrane, $C_{20}H_{20}$. All 20 protons are magnetically equivalent.

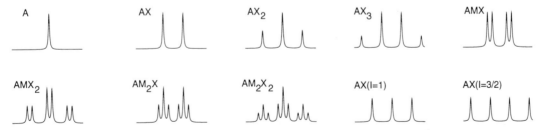

Fig. 3.13 Multiplet patterns for the A nucleus in various spin systems. M and X are spin-$\frac{1}{2}$ unless otherwise stated. Weak coupling is assumed throughout. Spectra are drawn for $|J_{AX}| > |J_{AM}|$.

3.3 Examples

Figure 3.14 shows the very different ^{31}P NMR spectra of three closely related phosphorus–sulphur compounds: αP_4S_4, βP_4S_4, and βP_4S_5. The multiplet structure arises entirely from ^{31}P–^{31}P couplings, because ^{32}S, the

Fig. 3.14 ^{31}P NMR multiplets of αP_4S_4, βP_4S_5 and βP_4S_4. The larger spheres represent the phosphorus atoms.

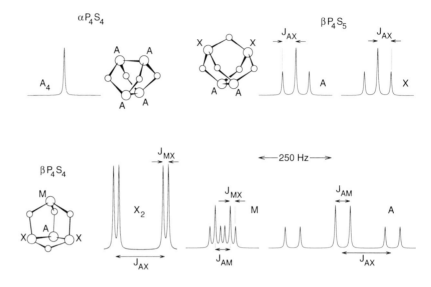

Fig. 3.15 ^{13}C NMR spectrum of $^6Li_4(^{13}CMe_3)_4$ and ^7Li NMR spectrum of $^7Li_4(^{13}CMe_3)_4$. Both spectra were recorded with proton-decoupling, a technique that removes the multiplet splittings caused by ^{1}H nuclei.

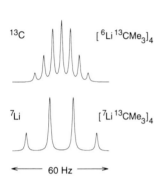

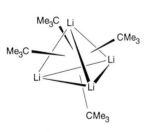

only isotope of sulphur with an appreciable natural abundance (99.24%), has spin $I = 0$. The three spin systems A_4 (αP_4S_4), AMX_2 (βP_4S_4), and A_2X_2 (βP_4S_5) are easily deduced from the spectra, and are clearly consistent with the structures shown.

The tetrameric structure of *t*-butyl lithium is clearly revealed by low temperature ^{13}C and ^{7}Li NMR (Fig. 3.15). The ^{7}Li spectrum of $[^7Li^{13}CMe_3]_4$ consists of a $1:3:3:1$ quartet: each lithium interacts with three equivalent *t*-butyl carbons, and has an unresolved (i.e. very small) coupling to the fourth, more distant ^{13}C. Similarly, the ^{13}C spectrum of $^6Li^{13}C(CH_3)_3$ is a septet, with relative intensities $1:3:6:7:6:3:1$, produced by each of the four equivalent ^{13}Cs interacting with three equivalent $I = 1$ ^{6}Li nuclei. The two coupling constants, $J(^7Li^{13}C) = 14.3$ Hz and $J(^6Li^{13}C) = 5.4$ Hz, are in the ratio of the gyromagnetic ratios of the two Li isotopes (1.04×10^8 and 3.94×10^7 T^{-1}s^{-1} respectively). As discussed in Chapter 4, these spectra are modified at higher temperatures by rapid rearrangement of the *t*-butyl groups.

A slightly more complex case is the ^{1}H spectrum of *meta*-bromonitrobenzene, Fig. 3.16. This is a weakly coupled AMPX spin system with all six pairwise couplings resolved, so that each proton gives a doublet of doublets of doublets, i.e. eight lines. The exact appearance of each multiplet is determined by the magnitudes of the coupling constants, and may readily be understood using $|J_{ortho}| > |J_{meta}| > |J_{para}|$. For the A and X multiplets, the

Fig. 3.16 ^{1}H NMR spectrum of *meta*-bromonitrobenzene. The six coupling constants are: $J_{AM} = 7.98$ Hz; $J_{AP} = 8.28$ Hz; $J_{MP} = 0.99$ Hz; $J_{MX} = 1.89$ Hz, $J_{PX} = 2.18$ Hz, $J_{AX} = 0.34$ Hz. Note that the couplings are in the order *ortho* > *meta* > *para*.

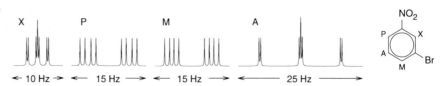

central pair of lines overlap strongly and appear as a single line of double intensity. Two further illustrations of the use of spin–spin couplings are given in Chapter 6 (Figs 6.19 and 6.20, p 86).

As an NMR nucleus, ^{13}C is second in popularity only to the proton: it is therefore appropriate at this point to comment briefly on multiplet splittings in ^{13}C spectra. In organic molecules, the dominant couplings experienced by ^{13}C nuclei are with their directly bonded protons. One-bond carbon–hydrogen coupling constants generally fall in the range 100–250 Hz, and are often an order of magnitude larger than two-bond and three-bond interactions. The multiplets produced by these couplings (a quartet for a methyl carbon, a triplet for a methylene, a doublet for a methine, and a singlet for a quaternary carbon) provide valuable clues when attempting to assign peaks in a spectrum to particular carbons in the molecule. However, ^{13}C NMR spectra are normally measured with the protons *decoupled*, so as to remove the ^{13}C–^{1}H splittings. This is achieved by irradiating the sample at the ^{1}H resonance frequency (about four times that of ^{13}C) while the ^{13}C spectrum is being recorded. The result is a considerably simplified spectrum: in the absence of heteronuclei (^{19}F, ^{31}P, etc.) each inequivalent carbon site in a molecule gives rise to a singlet in the proton-decoupled ^{13}C spectrum (denoted ^{13}C{^{1}H}).

Not only are ^{1}H-decoupled ^{13}C spectra less crowded than those with the proton–carbon couplings present, they also have much higher sensitivity. The latter arises from the nuclear Overhauser enhancement (a relaxation phenomenon described in Section 5.4) *and* because all the available NMR intensity for each ^{13}C is concentrated into a single line rather than being spread over several multiplet components.

Finally, *homonuclear* couplings are usually not observed in ^{13}C spectra because of the low natural abundance of ^{13}C (1.1%). Taking ethanol as an example, it is clear that of the molecules containing a ^{13}C at a given position, only about 1 in 100 contains a second ^{13}C. Thus, the spectrum of ^{13}CH$_3$^{13}CH$_2$OH should be about 100 times weaker than that of either ^{12}CH$_3$^{13}CH$_2$OH or ^{13}CH$_3$^{12}CH$_2$OH. ^{13}C–^{13}C splittings therefore often go unnoticed. ^{12}CH$_3$^{12}CH$_2$OH, by far the most abundant isotopomer, of course has no ^{13}C NMR spectrum at all.

For more on ^{13}C NMR see Wehrli *et al.* (1988).

3.4 Equivalent nuclei

We saw above that scalar couplings within groups of magnetically equivalent nuclei do not lead to multiplet splittings. To shed some light on this, let us consider initially the energy levels of two spin-$\frac{1}{2}$ nuclei, A and B, with identical chemical shifts and *no* spin–spin coupling (Fig. 3.17(a)). Denoting the $m = +\frac{1}{2}$ and $-\frac{1}{2}$ states of each spin α and β respectively, the four nuclear spin wavefunctions of the coupled pair are $\alpha_A\alpha_B$, $\alpha_A\beta_B$, $\beta_A\alpha_B$, $\beta_A\beta_B$. There are four allowed transitions, in which one of the spins flips independently of the other: $\alpha_A\alpha_B \leftrightarrow \alpha_A\beta_B$, $\alpha_A\alpha_B \leftrightarrow \beta_A\alpha_B$, $\alpha_A\beta_B \leftrightarrow \beta_A\beta_B$, $\beta_A\alpha_B \leftrightarrow \beta_A\beta_B$. When δν (the difference in resonance frequencies) and J (the coupling

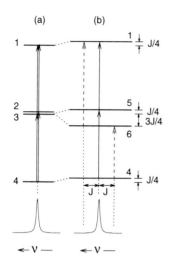

Fig. 3.17 Energy levels and spectra of a pair of *equivalent* spins A and B, with (b) and without (a) a mutual scalar coupling ($J = J_{AB} > 0$). Forbidden transitions are indicated by dashed vertical lines. Note that the spectrum is not affected by the presence of the scalar coupling. For clarity, the energy-level shifts due to scalar coupling have been greatly exaggerated. The spin states are as follows: 1, $\beta_A\beta_B$; 2, $\beta_A\alpha_B$; 3, $\alpha_A\beta_B$; 4, $\alpha_A\alpha_B$; 5, $2^{-1/2}(\alpha_A\beta_B + \beta_A\alpha_B)$; 6, $2^{-1/2}(\alpha_A\beta_B - \beta_A\alpha_B)$.

constant) are both zero, the four transitions are clearly degenerate, and the NMR spectrum consists of a single line at the chemical shift.

Now consider what happens when the scalar coupling is 'switched on'. If the spins were weakly coupled (i.e. $\delta v \gg J$), then the energy levels would shift as shown in Fig. 3.4(c): the upper and lower states, $\alpha_A\alpha_B$ and $\beta_A\beta_B$, would rise in energy by $\frac{1}{4}J$ (in frequency units, taking $J > 0$), while the central pair, $\alpha_A\beta_B$ and $\beta_A\alpha_B$, would sink in energy by the same amount. But, of course, equivalent spins ($\delta v = 0$) are not weakly coupled, and we should not expect identical behaviour. As indicated in Fig. 3.17(b), the upper and lower states are still destabilized by $\frac{1}{4}J$; however the central pair are *split apart* by $\pm\frac{1}{2}J$, which, on top of the expected shift of $-\frac{1}{4}J$, gives states with energies $+\frac{1}{4}J$ and $-\frac{3}{4}J$. This happens because the scalar coupling *mixes* $\alpha_A\beta_B$ and $\beta_A\alpha_B$ to produce non-degenerate states whose wavefunctions are $2^{-1/2}(\alpha_A\beta_B \pm \beta_A\alpha_B)$. Mixing of $\alpha_A\beta_B$ and $\beta_A\alpha_B$ does not occur for weakly coupled spins because the energy gap between the states (δv) is much larger than their interaction (J).

Now let us look more closely at the four wavefunctions of the two equivalent nuclei. The state described by the wavefunction $2^{-1/2}(\alpha_A\beta_B - \beta_A\alpha_B)$ is a *singlet* (the wavefunction is *antisymmetric*, i.e. it changes sign when the nuclear spin labels A and B are exchanged) and has no net magnetism because the magnetic moments of the two spins exactly cancel. The other three states $\alpha_A\alpha_B$, $2^{-1/2}(\alpha_A\beta_B + \beta_A\alpha_B)$ and $\beta_A\beta_B$ are all symmetric—the wavefunctions do *not* change sign when the nuclei are interchanged—and form the three sublevels of a *triplet*. (Parallel behaviour can be seen for pairs of *electrons*, for example the singlet and triplet states of atomic helium.) The three triplet energy levels are equally spaced (Fig. 3.17(b)) and can be thought of as arising from a 'compound' nucleus with spin-1; similarly, the singlet energy level can be regarded as that of a non-magnetic nucleus, $I = 0$. Looked at in this way, the NMR spectrum of two equivalent spins should simply be that of an isolated spin-1 nucleus, i.e. a single line at the chemical shift. Put another way, the two transitions involving the triplet energy levels, $\alpha_A\alpha_B \leftrightarrow 2^{-1/2}(\alpha_A\beta_B + \beta_A\alpha_B) \leftrightarrow \beta_A\beta_B$, are allowed and degenerate, while the singlet–triplet transitions, $\alpha_A\alpha_B \leftrightarrow 2^{-1/2}(\alpha_A\beta_B - \beta_A\alpha_B) \leftrightarrow \beta_A\beta_B$, which would have frequencies $\pm J$ either side of the chemical shift, are completely forbidden and have zero intensity (Fig. 3.17(b)).

To summarize, scalar couplings between magnetically equivalent nuclei do not produce multiplet splittings because of the changes in transition probabilities and NMR frequencies arising from the mixing of spin states by the spin–spin interaction. The same principles apply to larger numbers of equivalent spins.

3.5   Strong coupling

We have now looked at the two extremes of spin–spin coupling: $|\delta v| \gg |J|$ (weak coupling, Sections 3.1 and 3.2) and $\delta v = 0$ (equivalent nuclei, Section 3.4). Figure 3.18 shows what happens in between, when the spins are strongly coupled. A series of spectra for a range of δv values between $16J$

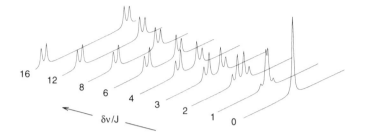

Fig. 3.18 Calculated NMR spectra of a pair of spin-$\frac{1}{2}$ nuclei for a range of δv values between $16J$ and zero.

and zero have been calculated for a pair of spin-$\frac{1}{2}$ nuclei. As the difference in resonance frequencies is reduced, keeping J fixed, the inner component of each doublet steadily increases in amplitude, while the outer components become weaker. The two doublets move together as their chemical shift difference decreases, until at $\delta v = 0$, the inner lines coincide and the outer lines vanish.

The origin of these changes may be understood by reference to Fig. 3.19. The effect of the scalar coupling is to *mix* the states $\alpha_A\beta_B$ and $\beta_A\alpha_B$ so modifying their wavefunctions and energies. The result is a change in the *transition probabilities* of the four NMR lines, the inner lines becoming more allowed (i.e. stronger) and the outer pair less allowed (weaker), the effect being more pronounced the smaller the gap δv between $\alpha_A\beta_B$ and $\beta_A\alpha_B$ relative to J. In the limit $\delta v = 0$, the outer lines are forbidden, as discussed in the previous section. According to convention, the two spectra in Fig. 3.19 are referred to as AX (weak coupling) and AB (strong coupling). The intensity distortions in the strongly coupled spectrum are sometimes referred to as the 'roof effect'. In the presence of strong coupling, the separation of the two central states is increased from δv to $C = [(\delta v)^2 + J^2]^{1/2}$, so that the doublets are no longer centred at the chemical shift positions.

As one might anticipate, the effects of strong coupling can be much more complicated when three or more spins are involved. The multiplet patterns discussed in Section 3.2 can be so severely distorted that they become difficult to recognize; the changes in transition probabilities cause otherwise forbidden transitions to be observed, and chemical shifts and coupling constants can no longer be extracted without detailed computer analysis. Fortunately, such problems are now largely a thing of the past. Because coupling constants are independent of the static magnetic field (Section 3.7) while δv is proportional to B_0 (eqn 2.3), a strongly coupled spin system can often be made weakly coupled by using a higher field spectrometer. For example, a pair of protons with $J = 5$ Hz would show an appreciable roof effect on a 60 MHz spectrometer if their chemical shift difference were less than about 1 ppm (i.e. $\delta v \leq 60$ Hz). At 600 MHz, the chemical shift difference would need to be less than 0.1 ppm to get a similar distortion. Statistically, strong coupling effects should thus be an order of magnitude less common at the higher field strength.

This section and the preceding one have given a 'hand-waving' account of the quantum mechanics of scalar-coupled spins; for the full treatment,

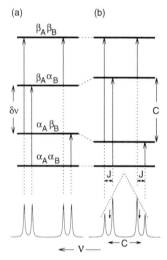

Fig. 3.19 Energy levels and spectra of a pair of spin-$\frac{1}{2}$ nuclei A and B, with strong coupling effects ignored (a) and included (b). The spin–spin interaction mixes states $\alpha_A\beta_B$ and $\beta_A\alpha_B$, altering their wavefunctions and energies. For clarity, the energy level shifts due to scalar coupling have been greatly exaggerated. $C = [(\delta v)^2 + J^2]^{1/2}$ where δv is the difference in the resonance frequencies of the two spins, and J is the scalar coupling constant. In (b), the 'roof effect' intensity distortion is indicated by the inverted 'V'; the arrows ($\downarrow$), which mark the chemical shifts of the two nuclei, are separated by δv.

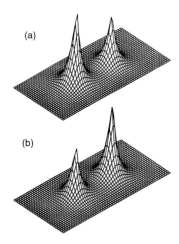

Fig. 3.22 Spin-polarized molecular orbitals of H_2 resulting from contact interactions. The sketches show the probability of finding an electron in (a) its α spin state and (b) its β spin state. The degree of spin polarization has been greatly exaggerated. (a) and (b) represent $|\phi_0 + \lambda\phi_1|^2$ and $|\phi_0 - \lambda\phi_1|^2$, respectively (see text and eqn 3.7).

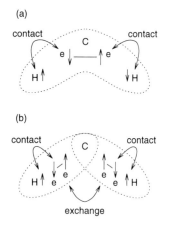

Fig. 3.23 Two simple models of scalar coupling in a CH_2 group: (a) a delocalized molecular orbital; (b) interacting localized molecular orbitals. In both cases, the most stable configuration of nuclear and electron spins is shown.

where λ is a small constant determined by the strength of the contact interaction and the energy of the excited state Ψ_1. Since ϕ_0 and ϕ_1 have different shapes, the probability of finding, say, electron a in state α at a given position in the molecule ($\approx|\phi_0 + \lambda\phi_1|^2$) is no longer identical to the probability for the same electron with the opposite spin, β ($\approx|\phi_0 - \lambda\phi_1|^2$): the electronic wavefunction has become *spin-polarized*, as shown in Fig. 3.22.

If proton A has $m = -\frac{1}{2}$ (β_A), the contact interaction and mixing of wavefunctions just outlined leads to an excess of α electron spins in its vicinity and a depletion of β spins (remember that the contact interaction stabilizes antiparallel electron and proton spin configurations). There is a corresponding build-up of β electron spins and reduction of α spins at the other proton B. So if B has $m = +\frac{1}{2}$ (α_B), it is stabilized by this local excess of β electrons through its contact interaction (Fig. 3.22). Conversely, if B has $m = -\frac{1}{2}$ (β_B), it is destabilized. In this way one nucleus senses the spin of the other. If proton A is inverted, to $m = +\frac{1}{2}$ (α_A), the situation is reversed, and there is a small accumulation of α electron spins around proton B, which is consequently destabilized in its α state.

This simple model of spin–spin coupling can, at least in principle, be extended to larger molecules. Consider for example a CH_2 group. If the electrons were completely localized in independent two-centre, two-electron CH bonds, it would be impossible for the contact interaction at one proton to polarize the electrons near the other proton. Clearly, the spins of the electrons involved in forming the two bonds must be correlated to some extent. This requirement can be visualized in two ways.

Using the same argument as for H_2, one can see that the protons will be coupled provided there exist *delocalized* molecular orbitals with appreciable s-electron density at the two protons, Fig. 3.23(a). Alternatively, the coupling may be pictured in terms of *interacting* two-centre, two-electron bonds, Fig. 3.23(b). The argument runs like this. (i) The spins of the electrons in one of the CH bonds are polarized via their contact interaction with the H nucleus. (ii) This polarization is transferred to the other CH bond by the exchange interaction (the electrostatic interaction that favours parallel electron spins over antiparallel) between electrons in the two bonds. (iii) The spin polarization of the second bond is felt by the second proton via its contact interaction.

Notice that these two pictures (Fig. 3.23) predict opposite signs for the coupling constant J_{HH}: positive in the first case (antiparallel nuclear spins stabilized) and negative in the second (parallel spins stabilized). In fact the coupling is rather more complex than either of these simple-minded models suggest. The signs of coupling constants cannot in general be forecast without detailed knowledge of the low-lying excited state wavefunctions. However, since the appearance of NMR spectra is usually independent of the signs of the coupling constants, this is not a serious problem.

It should be clear that similar arguments can be used to rationalize the existence of spin–spin interactions in larger molecules, and that the strength of the coupling is likely to fall off quite quickly as the number of intervening bonds increases.

The description of spin–spin coupling presented here is a simplified version of one to be found in Carrington and McLachlan (1967).

3.7 Properties of scalar coupling

The highly simplified arguments of the previous section give an impression of the mechanism of spin–spin coupling and indicate its general properties. The strength of the interaction is crucially dependent on the s-electron character of the ground state and electronically excited state wavefunctions at the positions of the nuclei. The coupling is not affected either by the strength or the direction of the external magnetic field, in contrast to the differences in resonance frequencies that arise from nuclear shielding. Indirect spin–spin couplings are therefore independent of the spectrometer frequency and, being isotropic, are not affected by molecular tumbling.

Calculation of spin–spin coupling constants is a non-trivial task requiring accurate wavefunctions for a large number of excited states. However it is easy to make an order of magnitude estimate of J_{HH}, the proton–proton coupling constant in H_2. It follows from the argument above that the value of J_{HH} in the (paramagnetic) *excited* state of H_2 (Ψ_1 in eqn 3.6) should be roughly equal to A_H, the contact interaction in atomic hydrogen. But we want J_{HH} for the *ground* state, Ψ_0, into which a small fraction (λ in eqn 3.7) of Ψ_1 has been mixed by the contact interaction. Simple perturbation theory shows that λ is of the order of $A_H/\Delta E$, where ΔE is the electronic excitation energy: J_{HH} is therefore approximately its value in state Ψ_1 scaled by λ, i.e. $J_{HH} \approx A_H^2/\Delta E$. Using $A_H \approx 10^9$ Hz (the hyperfine coupling constant for atomic hydrogen) and $\Delta E \approx 3 \times 10^{15}$ Hz (the singlet–triplet energy gap for H_2) one finds $J_{HH} \approx 300$ Hz, which, given the nature of the approximations involved, is surprisingly close to the actual value of 280 Hz (determined from the coupling constant in the isotopomer of hydrogen, H–D, using $J_{HH} = J_{HD}\gamma_H/\gamma_D$). Most coupling constants are smaller than 280 Hz, because the interaction occurs through more than one chemical bond and/or because one or both of the nuclei involved have smaller gyromagnetic ratios than ^{1}H.

One-bond and two-bond couplings

The interpretation of the magnitudes of scalar coupling constants is, in most cases, even more of a problem than it is for chemical shifts, and not one that will be tackled here. Instead, a few representative coupling constants are summarized (Figs 3.24, 3.25, 3.27, and 3.30) together with the briefest of comments.

One-bond carbon–proton couplings ($^1J_{CH}$) generally fall in the range 100–250 Hz, and are sensitive to the s-electron character of the carbon atomic orbital involved in the CH bond, reflecting the crucial role played by the contact interaction. The hydrocarbons ethane, ethylene and acetylene, which have respectively sp^3, sp^2 and sp hybridization, obey the empirical relation:

H_3C-CH_3	125	CH_4	125	CH_3Cl	147
$H_2C=CH_2$	157	CH_3OH	141	CH_2Cl_2	177
$HC\equiv CH$	250	CH_3CN	136	$CHCl_3$	208

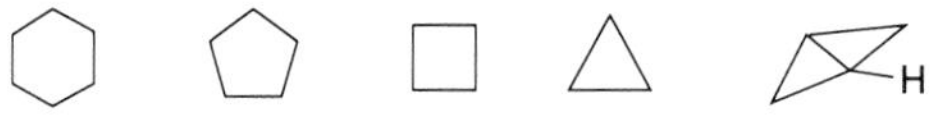

Fig. 3.24 One-bond $^{13}C-^1H$ coupling constants (in Hz).

123 128 136 161 205

$$^1J_{CH}/Hz \approx 5 \times \%(s) \tag{3.8}$$

where %(s) equals 25, 33 and 50 respectively (Fig. 3.24). Similar effects of hybridization are found for strained rings (Fig. 3.24): the smaller the ring size the larger the p-electron character of the C–C bonds in the ring, and consequently the larger the s-electron character of the carbon orbitals used to form the CH bonds. Figure 3.24 also gives a few examples illustrating the effect of substituents.

Two-bond (geminal) proton–proton couplings vary over a wide range (−23 to +42 Hz), with large substituent effects; sp^2 hybridized CH_2 groups generally have smaller $^2J_{HH}$ than do methyl groups (Fig. 3.25).

	$H_2C=CHX$	CH_3-X
X = H	+2.3	-12.4
X = Ph	+1.3	-14.5
X = Cl	-1.3	-10.8
X = CN	+0.9	-16.9

$H_2C=O$ +41

Fig. 3.25 Two-bond $^1H-^1H$ coupling constants (in Hz).

Three-bond couplings

Probably the most useful spin–spin couplings are those involving nuclei separated by *three* bonds, for example $^3J_{HH}$ in an H–C–C–H fragment. Experimentally and theoretically, these coupling constants are found to vary with the dihedral angle between the two H–C–C planes (θ, see Fig. 3.26) according to the 'Karplus relation':

$$^3J = A + B \cos\theta + C \cos^2\theta. \tag{3.9}$$

Although it is possible to calculate approximate values for A, B and C (including substituent and other effects), it is more satisfactory to treat them as coefficients to be determined empirically using conformationally rigid model compounds of known structure. Typical values are $A = 2$ Hz, $B = -1$ Hz, $C = 10$ Hz, which give a θ-variation of the type shown in Fig. 3.26 (a 'Karplus curve').

The utility of three-bond couplings lies principally in conformational analysis: $^3J_{HH}$ values for the ring protons in cyclohexanes depend on whether axial or equatorial protons are involved; and the *trans* proton–proton coupling constants across a C=C bond are up to a factor of two larger than the *cis* couplings (Fig. 3.27).

The Karplus relation finds valuable applications in studies of protein structures. For example, the couplings between the amide (NH) and C_α protons in a polypetide chain provide information on the conformation of the protein backbone (Fig. 3.28). In particular, the two major elements of secondary structure in proteins—helices and sheets—have characteristic H–N–C_α–H dihedral angles: ≈120 and ≈180° respectively. Thus, amide-C_α proton–proton coupling constants smaller than 6 Hz often indicate a helix

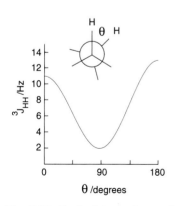

Fig. 3.26 Typical dependence of a three-bond H–C–C–H coupling constant on the dihedral angle θ, calculated using eqn 3.9.

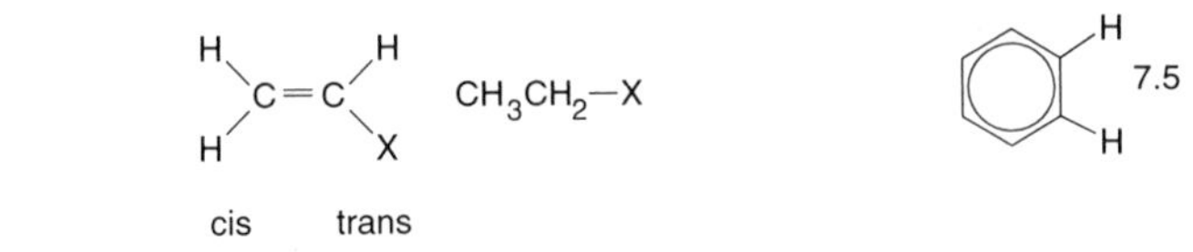

	cis	trans	
X = H	+11.5	+19.0	+8.0
X = Ph	+10.7	+17.5	+7.6
X = Cl	+7.4	+14.8	+7.2
X = CN	+11.8	+17.9	+7.6

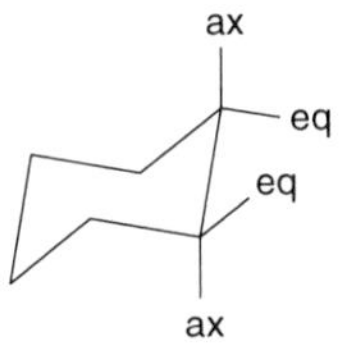

		θ
ax-ax	11.8	180°
ax-eq	3.9	60°
eq-eq	3.9	60°

Fig. 3.27 Three-bond ^{1}H–^{1}H coupling constants (in Hz).

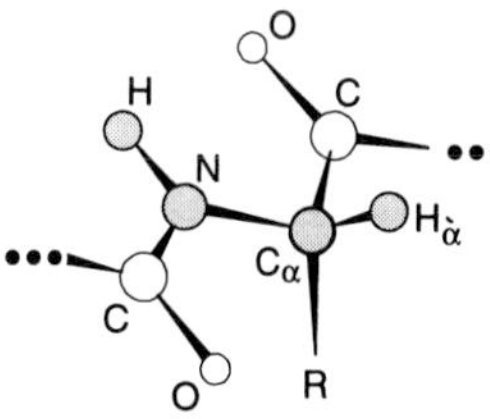

Fig. 3.28 Part of the backbone of a peptide chain, showing the H–N–C$_\alpha$–H dihedral angle. R is an amino acid side-chain.

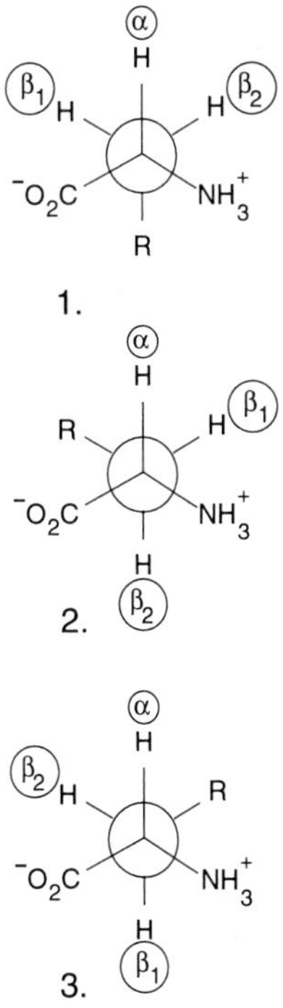

structure, while couplings larger than about 7 Hz generally arise from sections of the protein with a β-sheet structure.

The interpretation of three-bond couplings in conformationally mobile molecules is somewhat different. Consider, for example, the coupling between the α-proton and the two β-protons in an amino acid (Fig. 3.29). The three staggered conformations, or rotamers, interconvert rapidly so that the two observed $^3J_{\alpha\beta}$ values are averages, weighted according to the populations of the three energy minima (see Section 4.2 for an explanation of this averaging). Thus:

$$J_{\alpha\beta_1} = P_1 J_g + P_2 J_g + P_3 J_t$$
$$J_{\alpha\beta_2} = P_1 J_g + P_2 J_t + P_3 J_g \qquad (3.10)$$

where J_t and J_g are the *trans* ($\theta = 180°$) and *gauche* ($\theta = \pm 60°$) coupling constants, and $P_1 + P_2 + P_3 = 1$. The relative populations of the three rotamers can therefore be determined, provided J_t and J_g can be obtained from rigid model compounds.

Long-range couplings

Proton–proton coupling constants are generally very small (< 1 Hz) when the nuclei are separated by more than three bonds. A few of the exceptions are shown in Fig. 3.30. Note that large $^4J_{HH}$ and $^5J_{HH}$ often occur when the coupling is transmitted along a zigzag arrangement of bonds and/or through π-bonds.

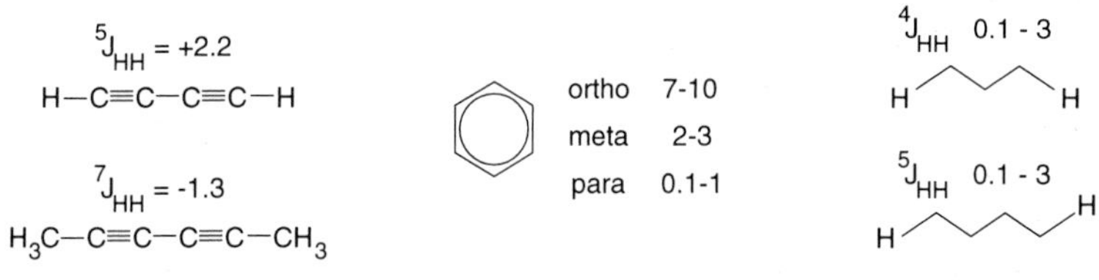

1.

2.

3.

Fig. 3.29 The three staggered conformations of an amino acid shown in Newman projection with C$_\alpha$ in front and C$_\beta$ behind.

$^5J_{HH} = +2.2$

H–C≡C–C≡C–H

$^7J_{HH} = -1.3$

H$_3$C–C≡C–C≡C–CH$_3$

	ortho	7-10
	meta	2-3
	para	0.1-1

$^4J_{HH}$ 0.1 - 3

$^5J_{HH}$ 0.1 - 3

Fig. 3.30 Long range ^{1}H–^{1}H coupling constants (in Hz).

3.8 Dipolar coupling

Finally, we turn to the direct dipolar interactions between nuclei which, though not responsible for splittings in the NMR spectra of liquids, are vital for an understanding of solid state NMR and relaxation processes.

Magnetic field of a point dipole

Every nucleus with non-zero spin quantum number has a *magnetic dipole*. A magnetic nucleus therefore behaves something like a small bar magnet with north and south poles, or a minute loop of wire carrying a current.

A classical magnetic moment $\boldsymbol{\mu}$, situated at the origin of an (x,y,z) coordinate system, and pointing along the positive z axis, produces a magnetic field $\boldsymbol{B_\mu}$ with cylindrical symmetry around the z axis. At a point (r,θ) in the $y = 0$ plane (r and θ are defined in Fig. 3.31) the components of $\boldsymbol{B_\mu}$ along the x, y and z axes are

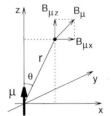

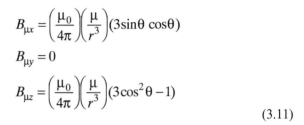

$$B_{\mu x} = \left(\frac{\mu_0}{4\pi}\right)\left(\frac{\mu}{r^3}\right)(3\sin\theta\,\cos\theta)$$

$$B_{\mu y} = 0$$

$$B_{\mu z} = \left(\frac{\mu_0}{4\pi}\right)\left(\frac{\mu}{r^3}\right)(3\cos^2\theta - 1)$$

$$(3.11)$$

Fig. 3.31 Parameters associated with the magnetic field generated by a magnetic dipole $\boldsymbol{\mu}$ at the origin of an (x,y,z) coordinate system. The components of the field $\boldsymbol{B_\mu}$ at a position in the $y = 0$ plane with polar coordinates (r,θ) are $(B_{\mu x}, 0, B_{\mu z})$, as specified in eqn 3.11.

where μ_0 is the permeability of vacuum ($4\pi \times 10^{-7}$ H m^{-1}), and μ is the magnitude of the magnetic moment $\boldsymbol{\mu}$. Note that $\boldsymbol{B_\mu}$ falls off as the *inverse cube* ($1/r^3$) of the distance from $\boldsymbol{\mu}$.

There are various ways of depicting the vector field $\boldsymbol{B_\mu}$. The most common, which we have already met in Chapter 2, is a plot of the *lines of force* (Fig. 3.32(a)), obtained by joining end to end infinitesimal vectors representing $\boldsymbol{B_\mu}$ (so that $\boldsymbol{B_\mu}$ is everywhere tangent to a line of force). Although this

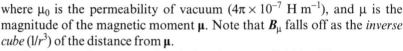

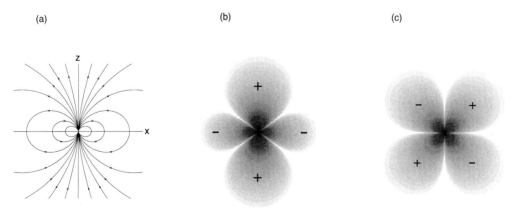

Fig. 3.32 Representations of the dipolar magnetic field $\boldsymbol{B_\mu}$ generated by a magnetic dipole $\boldsymbol{\mu}$ pointing along the positive z-axis: (a) lines of force; (b) $B_{\mu z}$, the z component of $\boldsymbol{B_\mu}$; (c) $B_{\mu x}$, the x component of $\boldsymbol{B_\mu}$. The positive and negative parts of $B_{\mu z}$ and $B_{\mu x}$ in (b) and (c) are indicated + and – respectively.

diagram gives little idea of the magnitude of B_μ, it does show its *direction* quite nicely. Thus, on the z axis ($\theta = 0$ or 180°), B_μ is parallel to μ; on the x axis (or indeed at any point in the $z = 0$ plane) B_μ is antiparallel to μ, i.e. it points along the *negative z* axis. At positions where $3 \cos^2\theta = 1$ ($\theta \approx 0 \pm 54.7°$ and $\theta \approx 180 \pm 54.7°$) the z component of the field vanishes, and B_μ is perpendicular to μ. These properties can of course also be seen directly from eqn 3.11. Thus the direction of B_μ depends on θ, while its magnitude is determined mainly by r, through the $1/r^3$ distance dependence.

A different way of visualizing B_μ is shown in Figs. 3.32(b) and (c) where its components parallel ($B_{\mu z}$) and perpendicular ($B_{\mu x}$) to the magnetic moment μ are drawn as density plots, which show clearly how B_μ changes as θ varies between 0 and 360°.

Note that if the magnetic moment μ is rotated, its magnetic field rotates with it; for example, if μ points along the negative z axis, B_μ is inverted.

Dipolar interaction between nuclei in solids

The effect one nucleus has on another may now be seen. Consider the X spectrum of an AX spin system, with X at (r, θ) and A at the origin. A and X are both spin-$\frac{1}{2}$ nuclei with *different* gyromagnetic ratios. In the strong magnetic field B_0 of an NMR spectrometer, both nuclear spins are quantized along the field direction (the z axis) so that the vertical axis in Fig. 3.32 coincides with B_0. The resonance frequency of X is determined by the net field in the z direction i.e. $B_0 \pm B_{\mu z}^A$, where $B_{\mu z}^A$ is the z component of the dipolar field generated by A (eqn 3.11) and the $\pm$ signs refer to the magnetic quantum number of A ($\pm\frac{1}{2}$). The splitting in the spectrum of X is therefore proportional to $B_{\mu z}^A$ and to the gyromagnetic ratio of X:

heteronuclear dipolar splitting/Hz $= K_{AX} (3 \cos^2\theta - 1)$

$$2\pi K_{AX} = \left(\frac{\mu_0}{4\pi}\right) \frac{\hbar \gamma_A \gamma_X}{r_{AX}^3}. \qquad (3.12)$$

For example, if the two nuclei are ^{1}H and ^{13}C, $K_{CH} = 8950$ Hz when $r_{CH} = 1.5$ Å; 472 Hz at 4 Å; and 30.2 Hz at 10 Å. Compared to scalar couplings, dipolar interactions are strong and long range.

Figure 3.33 shows NMR spectra calculated for a range of values of θ between 0 and 90°. These are the sort of spectra that would be observed for isolated AX pairs in a single crystal as the crystal is rotated in the magnetic field of the spectrometer. As θ changes from 0° ($3 \cos^2\theta - 1 = 2$) to 90° ($3 \cos^2\theta - 1 = -1$) the doublet splitting decreases, goes through zero at 54.7° (the so-called *magic angle*) and then increases again as θ rises to 90°. Identical behaviour is found in both the A and X spectra. Internuclear separations may easily be determined from single crystal spectra of such simple spin systems.

The situation is more complicated for a powdered sample. Although each AX pair has a unique value of θ, different molecules have different θ. Assuming a random distribution of orientations, the observed 'powder spectrum' is the sum of the single crystal spectra for θ between 0 and 90°,

Fig. 3.33 Calculated NMR spectra for one member of a heteronuclear pair (AX) of dipolar-coupled spin-$\frac{1}{2}$ nuclei. θ is the angle between the internuclear vector and the magnetic field direction. The dipolar coupling constant K_{AX} is given by eqn 3.12.

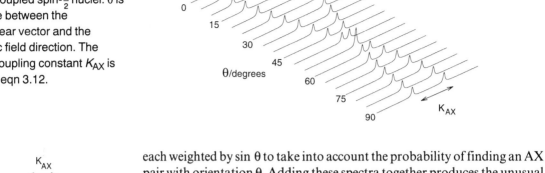

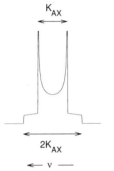

Fig. 3.34 Calculated powder spectrum for one member of a heteronuclear pair of dipolar-coupled spin-$\frac{1}{2}$ nuclei.

each weighted by $\sin\theta$ to take into account the probability of finding an AX pair with orientation θ. Adding these spectra together produces the unusual lineshape shown in Fig. 3.34: the 'horns' correspond to $\theta \approx 90°$, while the wings come from the $\theta \approx 0°$ orientations.

As may be anticipated, both single crystal and powder spectra are somewhat more complicated for larger spin systems, where every nucleus may have significant dipolar interactions with many neighbouring spins, each with its own r and θ.

Homonuclear dipolar couplings produce single crystal and powder spectra that are essentially identical to those from heteronuclear interactions. The only difference is that the expression for the observed splitting contains an extra factor of $\frac{3}{2}$:

$$\text{homonuclear dipolar splitting/Hz} = \tfrac{3}{2}\,K_{AX}(3\cos^2\theta - 1)$$

$$2\pi K_{AX} = \left(\frac{\mu_0}{4\pi}\right)\frac{\hbar\gamma^2}{r_{AX}^3} \tag{3.13}$$

Briefly, the $\frac{3}{2}$ arises because the homonuclear dipolar interaction mixes the spin states $\alpha_A\beta_X$ and $\beta_A\alpha_X$. This does not occur in the heteronuclear case because the gap between $\alpha_A\beta_X$ and $\beta_A\alpha_X$ (approximately $|\gamma_A - \gamma_X|B_0/2\pi$) is much greater than the strength of the interaction ($\approx K_{AX}$). The effect is very similar to the mixing produced by scalar couplings between equivalent and strongly coupled nuclei (Sections 3.4 and 3.5); however the anisotropy of the dipolar coupling changes the details of the mixing such that dipolar splittings *are* observed for equivalent nuclei. For example, the ^{1}H NMR spectrum of an isolated water molecule in a crystal is a doublet with splitting given by eqn 3.13 ($K_{HH} = 30.5$ kHz when $r_{HH} = 1.58$ Å).

Dipolar interaction between nuclei in liquids

Evidently dipolar couplings have a profound effect on the NMR spectra of solids. But what about liquids, with which this book is principally concerned?

Molecules in liquids rotate rapidly with frequent changes in the axis and speed of rotation as a result of collisions with other molecules.

Consequently, for any pair of nuclei, the angle θ and therefore the dipolar coupling become rapidly modulated. As discussed in more detail in Chapter 5, this leads to an *average* splitting, provided the frequency at which the coupling is modulated greatly exceeds the coupling itself. This condition is certainly met for all but very large molecules and/or very viscous solutions. For example, a water molecule at room temperature undergoes roughly 10^{12} revolutions per second, while the largest dipolar couplings are no more than 10^5 Hz. The angular part of eqns 3.12 and 3.13 must therefore be averaged over all orientations, θ:

$$\int_0^\pi (3\cos^2\theta - 1)\sin\theta \; d\theta = -\cos^3\theta + \cos\theta \Big|_0^\pi = 0 \qquad (3.14)$$

where, once again, $\sin\theta$ is the appropriate weighting for a molecule with no preferred orientation. This integral is identically zero: the positive parts of $(3\cos^2\theta - 1)\sin\theta$ exactly cancel the negative parts. Dipolar interactions do *not* therefore produce splittings in the NMR spectra of liquids: however they do play a crucial role in *spin relaxation*, as we shall see in Chapter 5.

The properties of scalar and dipolar couplings are summarized in Table 3.3.

Table 3.3 Properties of scalar and dipolar couplings

Scalar couplings	*Dipolar couplings*
Through bonds	Through space
Strength[a] $\leq$ 100 Hz	Strength[a] $\leq$ 100 kHz
Isotropic	$3\cos^2\theta - 1$
Small for > 3 bonds[a]	$1/r^3$
Produce splittings[b]	Splittings when motion restricted
Do not cause relaxation[c]	Cause relaxation when motion present

[a] Depends on gyromagnetic ratios of nuclei involved and molecular structure.
[b] Unless averaged by chemical exchange.
[c] Unless modulated by internal motion.

4   Chemical exchange

The two previous chapters have dealt with the interactions that give rise to peaks and splittings in NMR spectra. Let us now turn to processes capable of removing, or at least modifying, some of this structure, namely *dynamic equilibria*.

We start by considering the simplest case of dynamic equilibrium: a molecule converting between two conformations of equal energy:

$$A \underset{k}{\overset{k}{\rightleftharpoons}} B \qquad (4.1)$$

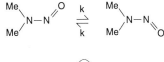

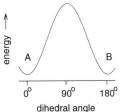

Fig. 4.1 Dimethylnitrosamine as an example of symmetrical two-site exchange. The two conformations of the molecule have identical energy and interconvert by 180° rotations of the nitroso group around the N–N bond, which has partial double-bond character. In the transition state, the N–NO plane is perpendicular to the C₂N–N skeleton of the molecule.

with identical forward and backward first-order rate constants, k. A good example is dimethylnitrosamine (Fig. 4.1). The skeleton of the molecule is planar, due to the partial double-bond character of the N–N bond, but not rigid. The nitroso group can undergo 180° rotations, so interconverting the two degenerate forms (Fig. 4.1). At low temperatures, the internal rotation is slow and the ^{1}H spectrum comprises two equally intense resonances from the methyl groups *cis* and *trans* to the oxygen (the coupling between the two sets of protons is too small to be resolved). Let the two resonance frequencies be ν_{cis} and ν_{trans}. At higher temperatures, the nitroso group flips at an appreciable rate, and every time it does so, the chemical shifts of the two sets of methyl protons are interchanged. The resonance frequency of each group of protons thus hops from ν_{cis} to ν_{trans} and back again, with an average time between jumps of $\tau = 1/k$. Processes such as this are known as *chemical exchange*, even though (as here) the dynamic equilibrium may not involve making and/or breaking of bonds.

What effect do such dynamic equilibria have on NMR spectra? The theory of chemical exchange is pleasingly straightforward compared to the dauntingly complex computations required for chemical shifts and coupling constants. Nevertheless, we shall avoid the algebra here (it can be found in almost all NMR textbooks, e.g. Carrington and McLachlan 1967; Harris 1983): the formulae involved are rather cumbersome, give little physical insight and are not essential for a qualitative appreciation of the origin of exchange effects and how they may be used to solve chemical problems. Rather, we summarize the results for a couple of simple cases, discuss their origin in handwaving terms and present a variety of examples and applications.

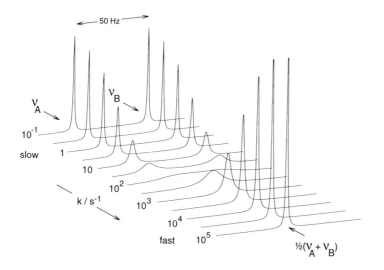

Fig. 4.2 Calculated NMR spectra for a pair of nuclei exchanging between two sites with equal populations (symmetrical two-site exchange). Spectra are shown for a range of values of the exchange rate k. The difference in resonance frequencies of the two sites, δv, is 50 Hz. The linewidths in the absence of exchange are 1 Hz.

4.1 Symmetrical two-site exchange

For the moment we stay with the simple two-site exchange $A \rightleftharpoons B$, with equal forward and backward rates. Figure 4.2 shows a set of spectra calculated for a range of rate constants, k. For very slow exchange, one sees two equally intense, narrow peaks at the resonance frequencies of the two sites, v_A and v_B. As k is increased, the two lines first broaden, then shift towards one another, broadening further until they merge into a single, wide, flat-topped resonance. This happens when k is similar in magnitude to the *difference* in resonance frequencies $|v_A - v_B|$ of the two exchanging sites. Further increase in the exchange rate produces a sharp resonance at the mean frequency $\frac{1}{2}(v_A + v_B)$. Thus, if the exchange is fast enough, the difference in resonance frequencies of the two sites collapses to zero. Experimentally, these changes may be observed by increasing the temperature.

Several questions immediately arise. Why are the two lines broadened by slow exchange? Why do they merge into a single sharp line when the exchange is fast, rather than continuing to get broader and broader as k increases? And why does the rate have to be large compared to $|v_A - v_B|$ to average the two environments? The following sections attempt to provide answers.

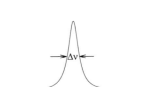

Fig. 4.3 Definition of Δv, the linewidth at half the maximum height of the resonance. In this chapter, Δv refers to the linebroadening arising from exchange processes, i.e. the total width less the width in the absence of exchange.

Slow exchange

Slow exchange is the regime in which the separate resonances are exchange-broadened but still to be found at frequencies v_A and v_B (see Fig. 4.2). In this limit, the increase in linewidth (defined in Fig. 4.3) as a result of exchange is simply given by

$$\Delta v = \frac{k}{\pi} = \frac{1}{\pi\tau}, \tag{4.2}$$

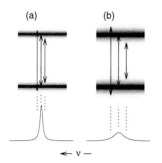

Fig. 4.4 Schematic representation of the blurring of energy levels and broadening of NMR lines brought about by the finite lifetime of spins undergoing chemical exchange, (b). In the absence of exchange (a), the linewidth is caused by spin relaxation process (see Chapter 5).

i.e. the faster the exchange, the wider the line. The origin of this effect is *lifetime broadening*, sometimes called *uncertainty broadening* because of its loose connection with the Heisenberg uncertainty principle. The energy of a state of finite lifetime cannot be specified precisely—the shorter lived the state, the greater the imprecision in its energy—and as indicated in Fig. 4.4, transitions between such 'blurred' energy levels result in broadened spectroscopic lines. It is for this reason that electronic transitions have larger natural linewidths than do rotational or vibrational spectra (faster spontaneous emission), and that microwave spectra broaden at high pressure (faster collisional deactivation of excited states).

Other contributions to NMR linewidths are discussed in Chapter 5.

Fast exchange

The other extreme, in which the two lines have merged to form a single, broadened resonance at the mean resonance frequency, is termed fast exchange. In this limit, the extra linewidth due to chemical exchange is:

$$\Delta\nu = \frac{\pi(\delta\nu)^2}{2k} = \tfrac{1}{2}\pi(\delta\nu)^2\tau \quad \text{where} \quad \delta\nu = \nu_A - \nu_B. \tag{4.3}$$

In contrast to slow exchange, where the separate resonances are broadened by site-hopping, the single line observed in fast exchange becomes *narrower* as the rate increases, reflecting the more effective averaging of the two environments A and B. That is, fast exchange causes the spins to experience an effective local field that is the mean of the two sites between which they hop.

To be able to detect *separate* signals from the two sites, they must acquire an appreciable *phase difference*, say 180°, which takes a time $\tfrac{1}{2}(\delta\nu)^{-1}$ (assume $\nu_A > \nu_B$), Fig. 4.5(a). Now, if exchange occurs during this period, the

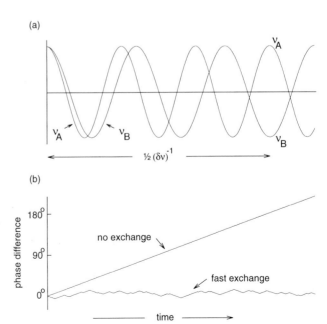

Fig. 4.5 (a) Accumulation of a phase difference between two spins with resonance frequencies ν_A and ν_B. (b) Time dependence of the phase difference in the absence of exchange, and under conditions of fast exchange.

accumulation of the phase difference is interrupted. When the spins swap frequencies, δv changes sign, and the phase difference starts to decrease. When the spins jump back to their original frequencies, their phase difference increases once more. So, as shown in Fig. 4.5(b), the phase difference undergoes a random walk with frequent reversals. The result, in the fast exchange regime, is a root-mean-square phase difference at the end of the $\frac{1}{2}(\delta v)^{-1}$ period that is much smaller than 180° so that, *in effect*, the two signals have very similar frequencies. In other words, fast exchange destroys the frequency difference $v_A - v_B$, provided the rate of site-hopping k is much faster than the build-up of the phase difference (rate $\approx \delta v$). The resonance observed in the fast exchange limit appears at the mean frequency $\frac{1}{2}(v_A + v_B)$, because each spin spends, on average, 50% of its time in each site.

Intermediate exchange

As its name implies, intermediate exchange fills the gap between slow and fast exchange: it is here that the separate resonances coalesce. The condition for the two resonances *just* to merge into a single broad line—the point at which the valley between the peaks is smoothed out—is

$$k = \frac{\pi \, \delta v}{\sqrt{2}} \approx 2.2 \, \delta v. \tag{4.4}$$

When k is *larger* than the right-hand side of this expression, a single line is expected at the mean resonance frequency; when k is *smaller*, two separate resonances should be seen (see Fig. 4.2).

NMR time-scale

Whether the exchange is slow, intermediate or fast is determined by the size of the exchange rate k relative to the frequency difference δv, as revealed by the discussion of phase differences above. This can also be seen from the condition for coalescence (eqn 4.4). That is, the *time-scale* of these events is governed by $(\delta v)^{-1}$. A process described as 'slow, or fast, on the NMR time-scale' is slow, or fast, *compared to the difference in resonance frequencies of the exchanging nuclei*. Since δv values are rarely larger than a few kHz, and often much smaller, only relatively slow equilibria (seconds to microseconds) can be studied by NMR.

For example, on a 100 MHz spectrometer, the resonances of two protons with chemical shifts separated by 1 ppm would merge when $k = 220$ s^{-1}, i.e. when the average lifetimes of the two sites are 4.5 ms. On a 500 MHz spectrometer, however, the exchange rate would need to be 1100 s^{-1} (910 μs lifetime) to cause the two lines to coalesce. Somewhat faster processes can often be studied by ^{13}C NMR because of the larger spread in resonance frequencies. Although the gyromagnetic ratio of ^{13}C is 25% of that of the proton, ^{13}C chemical shifts cover a ppm range about 20 times greater, giving, on average, resonance frequency differences larger by a factor of five.

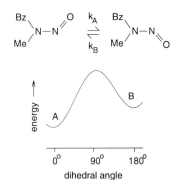

Fig. 4.6 Benzylmethylnitrosamine as an example of unsymmetrical two-site exchange. Bz ≡ C_6H_5–CH_2.

4.2 Unsymmetrical two-site exchange

Two-site exchange with equal populations is clearly a special case of the more general, and common, situation in which the two sites can have arbitrary concentrations. For example, if one of the methyl groups in dimethylnitrosamine is replaced by a benzyl group, the two conformers are no longer degenerate; the populations differ, as do the forward and backward rates (Fig. 4.6). Nevertheless, 180° flips of the nitroso group still modulate the chemical shifts of the methyl protons.

If the fractional populations of the two sites are p_A and p_B (i.e. $p_A + p_B = 1$) and the two first-order rate constants are k_A and k_B:

$$A \underset{k_B}{\overset{k_A}{\rightleftharpoons}} B \tag{4.5}$$

then $p_A k_A = p_B k_B$ at equilibrium. The average lifetimes of the two sites are $\tau_A = 1/k_A$ and $\tau_B = 1/k_B$.

In the slow exchange limit, the lifetime broadenings are

$$\Delta \nu_A = \frac{k_A}{\pi} = \frac{1}{\pi \tau_A} \quad \text{and} \quad \Delta \nu_B = \frac{k_B}{\pi} = \frac{1}{\pi \tau_B}. \tag{4.6}$$

Notice that the two resonances are unequally affected because of their different lifetimes (the less concentrated species has the shorter lifetime and the larger broadening). For fast exchange, a single resonance is observed at the *weighted* average frequency

$$\nu_{av} = p_A \nu_A + p_B \nu_B \tag{4.7}$$

with a linebroadening of

$$\Delta \nu = \frac{4\pi \, p_A p_B (\delta \nu)^2}{k_A + k_B}. \tag{4.8}$$

All these expressions reassuringly reduce to the formulae for symmetrical exchange when $p_A = p_B = \frac{1}{2}$, and $k_A = k_B = k$. Equation 4.7 should not be surprising: each nucleus spends a proportion p_A of its time in site A, and p_B in site B, giving an average resonance frequency weighted in favour of the more stable site. As a result, the exchange-averaged resonance moves around as the relative amounts of A and B and changed, for example by altering the temperature.

Figure 4.7 shows spectra calculated for a range of exchange rates for the case where B has twice the population of A. With increasing rate, one sees the same pattern of broadening, coalescence and subsequent narrowing as found for symmetrical exchange (Fig. 4.2).

The principles established for exchange between two sites may be applied to multiple-site exchange processes. The spectra are, not surprisingly, rather more complicated and generally need to be analysed by computer. Typically, an exchange pathway is proposed, spectra are calculated and the rate

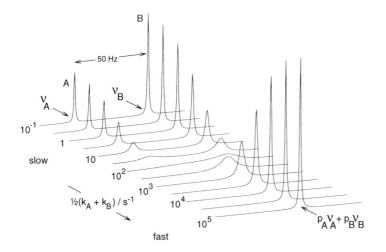

Fig. 4.7 Calculated NMR spectra for a pair of nuclei exchanging between two sites A and B with populations in the ratio $p_B/p_A = 2$ (unsymmetrical two-site exchange). Spectra are shown for a range of values of the average exchange rate $\frac{1}{2}(k_A + k_B)$, where $k_A/k_B = 2$. The difference in resonance frequencies of the two sites, δv, is 50 Hz. The linewidths in the absence of exchange are 1 Hz.

constant(s) varied to obtain a match between simulated and experimental spectra.

Processes other than exchange are capable of averaging resonance frequency differences. The most important instances are the averaging of dipolar (Section 3.8) and quadrupolar (Section 5.6) splittings by molecular tumbling in solution. The condition for averaging is the same as for fast exchange: the *modulation* of the frequency differences must be fast compared to the frequency differences themselves.

So far only *conformational* equilibria have been mentioned. As we shall see, similar and in some cases identical effects are found for *chemical* equilibria in which bonds are broken and formed. Other aspects of chemical exchange, and the variety of processes than can be studied, are best revealed with a few examples. Several instances of chemical exchange have already been encountered in previous chapters: e.g. fast exchange of the protonated and deprotonated forms of histidine (Section 2.1); the absence of certain splittings in the ^{1}H spectrum of ethanol (Section 3.2); the averaging of three-bond coupling constants by rapid internal rotation (Section 3.7).

4.3 Examples

A number of the following examples involve ^{13}C NMR spectra, recorded with proton decoupling (denoted ^{13}C$\{^1$H$\}$) at natural isotopic abundance of ^{13}C, so that each inequivalent carbon site gives rise to an NMR singlet, in the absence of other magnetic nuclei (see Section 3.3). ^{13}C$\{^1$H$\}$ NMR is particularly suited to the study of chemical exchange effects, whose theoretical analysis often becomes unnecessarily complicated in the presence of extraneous spin–spin couplings.

Cis-decalin—ring inversion

Cis-decalin, $C_{10}H_{18}$, consists of two cyclohexane rings, both in a chair conformation, fused *cis* to one another. Unlike the *trans* isomer, which is

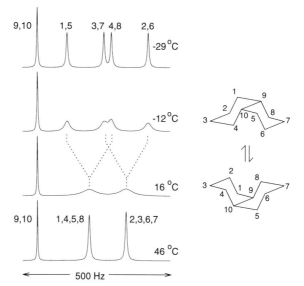

Fig. 4.8 $^{13}C\{^1H\}$ NMR spectra of *cis*-decalin as a function of temperature. (After D. K. Dalling, D. M. Grant, and L. F. Johnson, *J. Amer. Chem. Soc.*, 1971, **93**, 3678.)

conformationally rigid, *cis*-decalin flips between two degenerate conformations by chair-to-chair inversion of both rings, a process clearly revealed by $^{13}C\{^1H\}$ NMR (Fig. 4.8). At low temperature, when the inversion is slow, equally intense resonances are seen from the five pairs of equivalent carbons: (1,5), (2,6), (3,7), (4,8), and (9,10). Ring inversion at higher temperatures interchanges the chemical shifts as follows: 1⇌4, 5⇌8, 2⇌3, and 6⇌7, causing the (1,5) and (4,8) peaks to coalesce into a single line, and similarly the (2,6) and (3,7) resonances. The fifth ^{13}C signal, from carbons 9 and 10, is not affected, and remains sharp at all temperatures.

This simple example of symmetric two-site exchange may be analysed using the results of Section 4.1. Consider the (1,5) and (4,8) resonances, which have a chemical shift difference of 7.0 ppm, corresponding to a frequency difference δv of 175 Hz at the ^{13}C NMR frequency of 25 MHz. Rates of inversion, k, are easily determined as follows. At $-29°C$, the linebroadening from slow exchange (Δv) is 2.9 Hz, giving $k = 2.9\pi \approx 9$ s^{-1}. By $+27°C$, the two lines have coalesced to give a single line with an exchange broadening of 24.3 Hz. Using eqn 4.3, in the fast exchange limit, $k = \frac{1}{2}\pi (175)^2/24.3 \approx 2000$ s^{-1}. The (1,5) and (4,8) peaks coalesce at around $+5°C$, at which point $k = \pi (175)/\sqrt{2} \approx 400$ s^{-1}. Since the difference in chemical shifts of the two other pairs of exchanging carbons is almost the same as for (1,5) and (4,8), they coalesce at roughly the same temperature and have similar widths in fast exchange.

Bullvalene—degenerate rearrangement

Dynamic equilibria, such as the hindered internal rotation in *N*,*N*-dimethyl-nitrosamine (Section 4.1), which interconvert chemically identical forms of a molecule, are very conveniently studied by magnetic resonance. A

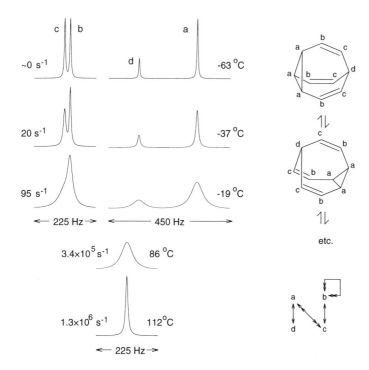

Fig. 4.9 $^{13}C\{^1H\}$ NMR spectra of bullvalene as a function of temperature. Approximate exchange rates are shown to the left of each spectrum. The exchange pathways are summarized at bottom right, the number of arrow heads corresponding to the number of sites exchanging. (After H. Günther and J. Ulmen, *Tetrahedron*, 1974, **30**, 3781.)

celebrated example is bullvalene, a molecule comprising ten C–H groups in a cage structure with a three-fold axis of symmetry (Fig. 4.9).

At −63°C in $CHCl_2$–$CHCl_2$ solution, the 22.5 MHz $^{13}C\{^1H\}$ spectrum consists of four resonances, from the four distinct carbon sites, with intensity ratios 3 : 3 : 3 : 1, reflecting the symmetry of the molecule. As the temperature is increased, all four broaden and coalesce into a single sharp peak at the weighted-average chemical shift, $(3\delta_a + 3\delta_b + 3\delta_c + \delta_d)/10$.

The process responsible is a series of degenerate Cope rearrangements, which permutes the carbons, but leaves the structure unaltered. Each molecule may rearrange in three ways by breaking a different cyclopropane bond, leading to a complex network of thousands of interconverting structures, with all possible permutations of C–H groups.

A detailed analysis using computer-simulated spectra yields the rate constants shown in Fig. 4.9. An Arrhenius plot (ln k against $1/T$) gives an activation energy $E_a \approx 60$ kJ mol^{-1} and pre-exponential factor $A \approx 10^{14}$ s^{-1}. Similar exchange effects are observed in the ^1H spectrum, but are more complicated to interpret because of the ^1H–^1H couplings.

Finally, it is interesting to note that resonance b is narrower in the slow exchange regime than the other three, which all have similar widths (e.g. at −37°C). The reason is that every time the molecule undergoes a rearrangement, only one of the three carbons in site b changes its resonance frequency (the other two stay in site b); all seven remaining carbons are moved to a different site (see structures in Fig. 4.9). The average lifetime of the carbons in site *b* is therefore three times longer, and the exchange broadening three times smaller than for sites *a*, *c* and *d*.

This differential linebroadening can be used to determine the mechanism of a rearrangement, as the following example illustrates.

$(\eta^4-C_8H_8)Ru(CO)_3$—mechanism of rearrangement

Below about −100°C, the eight carbons of the cyclooctatetraene ligand in $(\eta^4-C_8H_8)Ru(CO)_3$ give rise to four equally intense $^{13}C\{^1H\}$ resonances, consistent with a structure in which the ruthenium is attached to two of the four double bonds of the C_8H_8 ring (Fig. 4.10). (No ruthenium–carbon couplings are observed because of rapid quadrupolar relaxation (see Section 5.6) of both ^{99}Ru $(I = \frac{5}{2})$ and ^{101}Ru $(I = \frac{5}{2})$, the only magnetic isotopes of Ru.) As the temperature is increased, peaks *b* and *c* broaden more rapidly than *a* and *d*. Above about −60°C, a single ^{13}C resonance is observed at the average chemical shift of the four peaks indicating rapid rearrangement of the ligand, rendering all eight carbons effectively equivalent.

Figure 4.10 shows the four possible 1,*n* shifts (*n* = 2,3,4,5) of the Ru(CO)$_3$ group. A 1,5 shift can be excluded immediately: it simply swaps sites *a* and *d*, and sites *b* and *c*, and so would lead to *two* peaks at chemical shifts $\frac{1}{2}(\delta_a + \delta_d)$ and $\frac{1}{2}(\delta_b + \delta_c)$ in fast exchange.

Of the remaining rearrangement pathways, only a 1,2 shift is consistent with the low temperature spectra. As shown in Fig. 4.10, a 1,2 jump has no effect on the chemical shift of *one* of the carbons in each of sites *a* and *d*, but moves *both* of the *b* and the *c* carbons to another site with a different resonance frequency. On average, therefore, sites *a* and *d* have *twice* the

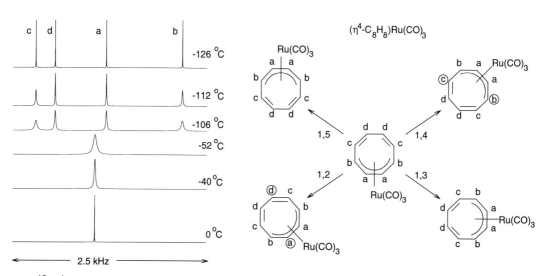

Fig. 4.10 $^{13}C\{^1H\}$ NMR spectra of the cyclooctatetraene carbons of $(\eta^4-C_8H_8)Ru(CO)_3$ as a function of temperature. The observed pattern of linebroadening is only compatible with a 1,2 rearrangement mechanism. (After F. A. Cotton and D. L. Hunter, *J. Amer. Chem. Soc.*, 1976, **98**, 1413.)

lifetime of *b* and *c* and show *half* the linebroadening. A 1,4 shift of the Ru(CO)₃ group has exactly the opposite effect on the four sites and would cause peaks *a* and *d* to have *twice* the width of *b* and *c*. By contrast, a 1,3 jump of the Ru(CO)₃ group alters the resonance frequency of all eight carbons and so would broaden all four peaks to the same extent; a random jump mechanism would have a similar effect.

[Li(*t*-Bu)]₄—averaging of spin–spin couplings

So far the impression has been given that the only NMR properties that can be averaged by fast exchange are chemical shifts. In fact the same principles apply to spin–spin couplings and relaxation times, although we shall not discuss the latter here.

In cyclopentane solution, *t*-butyllithium is tetrameric: the lithium atoms are placed at the corners of a tetrahedron with a C(CH₃)₃ group sitting above each of the triangular faces, to give a cubane-type structure. Consistent with this, the α-carbons of the *t*-butyl groups in the ⁶Li compound at temperatures below –22°C show a single ¹³C{¹H} resonance with seven equally spaced lines (Figs 4.11 and 3.15). Each ¹³C couples equally to three equidistant ⁶Li nuclei to give a 1 : 3 : 6 : 7 : 6 : 3 : 1 multiplet pattern (⁶Li has $I = 1$ and undergoes slow quadrupolar relaxation), with no resolvable coupling to the fourth, more remote, ⁶Li. The ⁶Li–¹³C coupling constant is 5.4 Hz.

At temperatures above –5°C, the septet is replaced by a nonet (nine lines) with splitting 4.1 Hz: rapid intramolecular scrambling of the *t*-butyl groups causes the ¹³C to interact equally with all four lithium nuclei, with an average coupling constant: $(3 \times 5.4 + 0)/4 \approx 4.1$ Hz.

A further example of the averaging of coupling constants is provided by liquid ethanol.

Ethanol—proton exchange

In Chapter 3 we saw how the interaction between the CH₃ and CH₂ protons in liquid ethanol leads to a triplet and a quartet in the ¹H spectrum. But what of the other three-bond coupling in the molecule, that between the OH and CH₂ protons? The splitting due to this interaction is only visible in highly purified ethanol, from which all traces of acid and base have been removed. Under these stringent conditions, the OH singlet splits into a 1 : 2 : 1 triplet ($J = 4.8$ Hz), and each line of the CH₂ quartet becomes a doublet with the same coupling constant (Fig. 4.12). On addition of as little as 10^{-5} mol dm⁻³ acid, these extra splittings vanish as a result of rapid acid-catalysed H⁺ exchange.

Consider an ethanol hydroxyl proton (A) exchanging with an H⁺ ion (B):

$$CH_3–CH_2–OH_A + H_B^+ \rightleftharpoons CH_3–CH_2–\overset{+}{O}\!\!\begin{smallmatrix}H_A\\ \\H_B\end{smallmatrix} \rightleftharpoons CH_3–CH_2–OH_B + H_A^+.$$

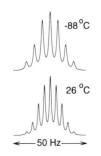

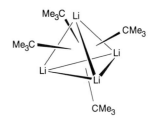

Fig. 4.11 ¹³C{¹H} NMR spectra of the *t*-butyl α carbon of [⁶Li¹³CMe₃]₄ in cyclopentane at two temperatures. The septet (–88°C) and nonet (26°C) are formed by coupling to respectively three and four equivalent $I = 1$ ⁶Li nuclei. (After R. D. Thomas, M. T. Clarke, R. M. Jensen, and T. C. Young, *Organometallics*, 1986, **5**, 1851.)

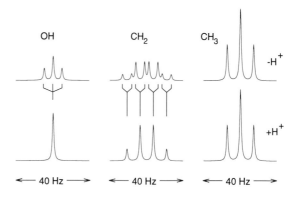

Fig. 4.12 ^{1}H NMR multiplets of ethanol with (below) and without (above) a trace of acid to catalyse intermolecular hydrogen transfer.

If H_A and H_B have the same magnetic quantum number ($m = \pm\frac{1}{2}$), the exchange of protons has no effect on the resonance frequency of the CH_2 protons, and is undetectable by NMR. However, when H_A and H_B have opposite spin orientations ($\uparrow$ and $\downarrow$), the net effect of the exchange is

$$CH_3-CH_2-OH\uparrow + H^+\downarrow \rightleftharpoons CH_3-CH_2-OH\downarrow + H^+\uparrow.$$

The ethyl protons cannot tell whether the hydroxyl proton is A or B, only whether it is $\uparrow$ or $\downarrow$. So, at each H^+ exchange, the CH_2 resonance frequency changes by $\pm J$. If this hopping is fast compared to the frequency difference involved, i.e. J, the doublet splitting of the CH_2 multiplet collapses to zero, in the same way that chemical shift differences are removed by fast exchange. Clearly, the triplet structure of the OH resonance also disappears when proton exchange is faster than about 10 Hz.

The influence of H^+ exchange is also evident in the ^{1}H spectrum of ethanol–water mixtures (Fig. 4.13). When acid is rigorously excluded, separate OH resonances for ethanol and water can be seen; addition of a little acid leads to fast exchange and a single peak at the average chemical shift, weighted by the relative concentrations of H_2O and ethanol OH protons.

Finally, there is one more exchange process in ethanol which has rather been taken for granted so far. This is the averaging, by rapid internal rotation around the C–C and C–O bonds, of the chemical shift differences of the three methyl protons, and of the two methylene protons, which arise from the various conformations of the molecule. At sufficiently low temperatures, when the rate of internal rotation is comparable to these frequency differences, the ^{1}H spectrum of this endlessly fascinating molecule would therefore become more complicated than Fig. 4.12.

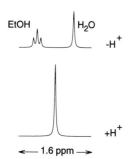

Fig. 4.13 ^{1}H spectra of the OH resonances of ethanol and water with (below) and without (above) a trace of acid.

Motion of tyrosine side-chains in proteins

Recent advances in NMR techniques have made it possible to determine the conformations of small proteins (up to about 20 000 molecular weight, see Section 5.7) in solution. Important though it is to know their structures, protein molecules are far from rigid, and information on their dynamical properties is essential for a complete understanding of their functions. A simple way in which NMR can provide such insights is through the exchange effects associated with the internal rotation of groups such as the phenol side-chain of the amino acid tyrosine.

In a rigid protein, the four ring protons of a tyrosine have different local environments and, barring coincidences, different chemical shifts. In particular, the asymmetric surroundings (Fig. 4.14) cause H2 and H6 to be inequivalent, and similarly H3 and H5. However, if the ring undergoes 180° flips around its two-fold axis, H2 and H6 swap chemical shifts, as do H3 and H5, and if this exchange is fast compared to the frequency differences, only two ^{1}H resonances will be observed, at positions $\frac{1}{2}(\delta_2 + \delta_6)$ and $\frac{1}{2}(\delta_3 + \delta_5)$.

The ring protons of a rigid tyrosine residue in a protein typically cover a chemical shift range of 1 ppm, so that coalescence of the four aromatic resonances occurs at a flipping rate of roughly 500 s^{-1} on a 500 MHz spectrometer, and lineshape changes are detectable for flipping rates between a few times per second up to about 10^5 s^{-1}. The rate at which an aromatic ring rotates is governed principally by the dynamics of neighbouring groups which must move aside to allow the ring to flip. The ^{1}H NMR spectra of tyrosine and phenylalanine (tyrosine without the hydroxyl group) side-chains thus give information on these relatively slow cooperative structural fluctuations, over a broad time-scale.

As an example, Fig. 4.14 shows a series of ^{1}H spectra of basic pancreatic trypsin inhibitor. BPTI is a small globular protein of molecular weight 6500, consisting of a single polypeptide chain of 58 amino acid residues of which four are tyrosines. Below about 30°C, Tyr-35 exhibits four separate aromatic resonances, which merge into two above 70°C. Similar behaviour is found for Tyr-23 (not shown), at a temperature some 30°C lower. Both Tyr-10 and Tyr-21 give two resonances, even at 4°C, and are essentially free rotors.

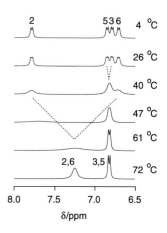

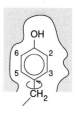

Fig. 4.14 ^{1}H NMR signals of Tyr-35 in basic pancreatic trypsin inhibitor, BPTI, as a function of temperature. The structure indicates the asymmetric environment of a tyrosine side-chain in a protein. The ring flipping rates are 2.5 × 10^{-2} s^{-1} at 4°C, and 1.7 × 10^4 s^{-1} at 72°C. (After K. Wüthrich, *NMR of Proteins and Nucleic Acids*, Wiley, New York, 1986.)

5 Spin relaxation

Chemical shifts, spin–spin couplings and chemical exchange are the factors that chiefly determine the appearance of an NMR spectrum. More subtle, but no less important, are the two relaxation processes—*spin–lattice relaxation* and *spin–spin relaxation*—which allow nuclear spins to return to equilibrium following some disturbance.

5.1 Spin–lattice relaxation

Imagine doing an NMR experiment. Before the sample is put into the magnet, the $2I+1$ energy levels of a spin–I nucleus are degenerate (neglecting the tiny splitting produced by the earth's magnetic field) and their populations are equal. When the sample is dropped into the magnetic field, the spin states suddenly split apart in energy. The populations, however, cannot adjust themselves instantaneously, and remain equal. If this non-equilibrium state were to persist, NMR spectroscopy would be impossible: fortunately, *spin–lattice relaxation* comes to the rescue by enabling the spins to flip amongst their energy levels so as to establish the Boltzmann population differences required for a successful NMR experiment. As the nuclei approach equilibrium, the energy released is dissipated in the surroundings (the 'lattice').

These changes in populations are characterized by a time T_1—*the spin–lattice relaxation time*. For a collection of spin-$\frac{1}{2}$ nuclei, assuming exponential relaxation (which is often the case), the difference in the number of $m = +\frac{1}{2}$ and $m = -\frac{1}{2}$ spins grows according to

$$\Delta n(t) = \Delta n_{\text{eq}}[1 - \exp(-t/T_1)] \qquad (5.1)$$

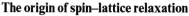

where Δn_{eq} is the equilibrium population difference and t is time (see Fig. 5.1). After putting the sample in the magnet, one should therefore wait a time long compared to T_1, before trying to record a spectrum. This is not normally a problem as typical T_1 values for spin-$\frac{1}{2}$ nuclei in liquids are no more than a few seconds.

Similar behaviour is to be expected whenever nuclear spin populations are disturbed. As we shall see in Section 6.3, there are better ways of measuring T_1 values than simply dropping the sample into a magnet and trying to watch the NMR signal appear.

The origin of spin–lattice relaxation

Relaxation mechanisms that operate in other forms of spectroscopy are ineffective for NMR. Spontaneous emission (fluorescence), whose rate depends on the frequency of the transition *cubed*, is exceedingly slow at NMR

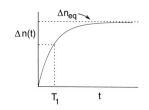

Fig. 5.1 The difference (Δn) in $m = +\frac{1}{2}$ and $m = -\frac{1}{2}$ populations of a collection of spin-$\frac{1}{2}$ nuclei as a function of the time (t) after putting the sample in a magnetic field, eqn 5.1. By the time $t = T_1$, where T_1 is the spin–lattice relaxation time, the population difference has grown to 63% of its value (Δn_{eq}) at thermal equilibrium.

frequencies. Deactivation of excited states by molecular collisions is also negligible because nuclear spins interact so weakly with the rest of the world that they are effectively decoupled from the motions of the molecules that contain them. As a molecule rotates, its nuclear spins remain aligned with the magnetic field, rather than reorienting with the molecule; in this respect a nuclear spin behaves like a ship's gimbal compass which points north however much the ship pitches and yaws.

The mechanism of nuclear spin relaxation lies, not surprisingly, in magnetic interactions, the most important being dipolar coupling. We saw in Section 3.8 that the dipolar coupling between two nuclei depends on their separation r, and on θ, the angle between the internuclear vector and the static field. Although this purely anisotropic coupling leads to no splittings in the NMR spectra of molecules in liquids, the *instantaneous* interaction is far from negligible. As the molecules translate, rotate and vibrate in solution, r and θ vary in a complicated way causing the interaction to fluctuate rapidly. Thus the dipolar coupling, modulated by molecular motions, causes nuclear spins to experience time-dependent local magnetic fields which, if they contain a component at the NMR frequency ($\approx \gamma B_0/2\pi$), can induce the radiationless transitions which return the spins to equilibrium. Most other relaxation mechanisms have essentially the same origin: an intramolecular or intermolecular magnetic (or, for $I > \frac{1}{2}$ nuclei, electronic) interaction, rendered time-dependent by random molecular motion.

Before saying more about spin–lattice relaxation, we need to look a little at the rotational motion of molecules in liquids.

5.2 Rotational motion in liquids

In gases, at least at low pressure, the mean free path is large and a molecule can rotate end-over-end many times before suffering a collision that changes its rotational state. In liquids, collisions occur much more frequently; molecules are buffeted constantly from all sides, each bump accelerating or decelerating their rotational motion and deflecting the axis of rotation. This motion is often referred to as *tumbling* rather than rotation, to reflect its chaotic nature.

Let us focus on a simple molecule, CH_4, and imagine the carbon atom fixed at the centre of a sphere with radius equal to the CH bond length. As the methane molecule collides with its neighbours in solution, each of the four hydrogen atoms undergoes a random walk on the surface of the sphere. Figure 5.2 shows one such zigzag trajectory for an H atom, initially at the 'north pole'. Every CH_4 undergoes a different random walk, determined by its collisions with other molecules. We can visualize this by plotting, for many such trajectories, the position reached by the H atom a certain time after setting off from the north pole (Fig. 5.3). After a short time (a), the molecules have spread out a bit on the sphere, but are still clustered around their starting position. Somewhat later (b), the average displacement has increased and some have reached the equator. After a longer time, there are substantial numbers of H atoms in the southern hemisphere but still an

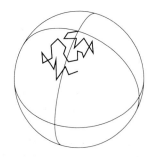

Fig. 5.2 A typical random walk trajectory for a molecule undergoing rotational diffusion in a liquid.

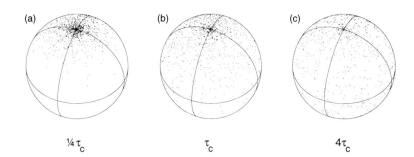

(a)   (b)   (c)

¼τ_c   τ_c   4τ_c

Fig. 5.3 Representative positions of molecules undergoing rotational diffusion in a liquid, at different times after starting off from the 'north pole': (a) $t = \tau_c/4$; (b) $t = \tau_c$; (c) $t = 4\tau_c$; where τ_c is the rotational correlation time.

excess at northern latitudes (c). If we were to wait a very long time, the molecular orientations would be randomly distributed around the sphere.

We can define a characteristic time for this motion, the *rotational correlation time*, τ_c. Roughly speaking, τ_c is the time taken for the root-mean-square deflection of the molecules to be about 1 radian ($\approx 60°$). At times much less than τ_c, most of the molecules are close to their original positions, while for $t \gg \tau_c$, the orientations have been completely randomized and all 'memory' of the original orientation is lost. Typical values for τ_c for small molecules in non-viscous solvents at room temperature are in the region of 100 ps.

The point of Figs 5.2 and 5.3 is that they allow us to see, qualitatively, what the frequency spectrum of this random motion ought to look like. Since τ_c is the average time taken to tumble through an angle of about 1 radian, τ_c^{-1} must be approximately the root-mean-square rotational frequency (in radians s^{-1}). Further, Fig. 5.3 shows that frequencies *less than* τ_c^{-1} will be reasonably probable, since these correspond to rotations of less than 1 radian during a time τ_c, while frequencies *above* τ_c^{-1}, which correspond to rotations greater than 1 radian, will be much less likely.

In short, the frequency spectrum of an intramolecular magnetic interaction, modulated by molecular tumbling, should resemble one of the curves drawn in Fig. 5.4. This function is given the symbol $J(\omega)$ and is called the *spectral density* (ω is the angular frequency in radians s^{-1}). It can be thought of as being proportional to the *probability* of finding a component of the random motion at a particular frequency. As such, the integral of $J(\omega)$ over all frequencies is a constant, independent of τ_c. The frequency dependence of $J(\omega)$ is governed by τ_c. Smaller molecules, or less viscous solvents, or higher temperatures should all result in shorter correlation times (faster tumbling, on average) and hence a spectral density that extends to higher frequencies, as shown in Fig. 5.4 for three different τ_c values. These curves have been drawn using the functional form:

$$J(\omega) = \frac{2\tau_c}{1 + \omega^2 \tau_c^2} \tag{5.2}$$

which is appropriate when the molecule's 'memory' of its orientation at an earlier time decays *exponentially*. For this particular spectral density, $J(\omega) = \frac{1}{2}J(0)$ when $\omega = \tau_c^{-1}$.

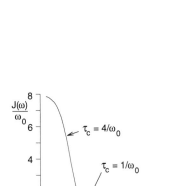

Fig. 5.4 The spectral density function $J(\omega)$ (eqn 5.2) drawn for three values of the rotational correlation time τ_c. ω_0 is the NMR frequency. The horizontal axis is logarithmic. Note that $J(\omega_0)$ has its maximum value when $\tau_c = 1/\omega_0$. As τ_c is reduced, $J(\omega)$ extends to higher frequencies and is generally flatter (because the area under the curve is fixed).

5.3 Spin–lattice relaxation again

Although the dipolar interaction is the most common source of relaxation, it is not the simplest. Two dipolar-coupled nuclei experience *correlated* time-dependent magnetic fields (they have the same r and θ in eqn 3.13) and consequently the two spins relax in a concerted manner. Although this is the source of some interesting and useful relaxation effects (Section 5.4), it is an unwelcome complication at this stage. To keep things simple, the following paragraphs discuss an idealized mechanism in which nuclear spins are independently relaxed by random local fields. The conclusions drawn give insight into spin relaxation in general, and only require a slight adjustment to yield quantitative predictions for particular relaxation mechanisms.

Spin–lattice relaxation is caused by fluctuating local fields which induce nuclei to flip amongst their available spin states. The rate of this process, T_1^{-1}, depends on the probability that the local fields have a component oscillating at the appropriate frequency, namely the NMR frequency, $\omega_0 \approx \gamma B_0$. In other words T_1^{-1} is proportional to the spectral density $J(\omega_0)$.

Following the last section, we have the information necessary to predict how the spin–lattice relaxation rate varies with the rate of molecular motion. Figure 5.4 shows that $J(\omega_0)$ is small for root-mean-square tumbling rates (i.e. τ_c^{-1}) much smaller than ω_0, or much larger than ω_0, and reaches a maximum when τ_c^{-1} matches the resonance frequency (i.e. $\omega_0\tau_c = 1$). This behaviour is summarized in Fig. 5.5. For rapidly tumbling molecules with $\omega_0\tau_c \ll 1$ (left-hand side of Fig. 5.5), $J(\omega_0) \approx 2\tau_c$ and the relaxation gets *slower* as the mean tumbling rate is increased (e.g. by raising the temperature). Conversely, slowly tumbling molecules have $\omega_0\tau_c \gg 1$ (right-hand side) and $J(\omega_0) \approx 2/\omega_0^2\tau_c$, so that the relaxation *accelerates* as the tumbling speeds up. The maximum relaxation rate (minimum T_1) occurs for $\omega_0\tau_c = 1$, at which point $J(\omega_0) = 1/\omega_0$.

To decide whether a given molecule falls on the left or the right of the T_1 minimum, one clearly needs to know τ_c. This information is often not available—indeed τ_c values are often obtained from relaxation time measurements—but a rough and ready rule of thumb can be given (Sanders and Hunter 1993). For molecules in water at room temperature, τ_c is very approximately related to the relative molecular mass M_r by

$$\tau_c/\text{ps} \approx M_r. \tag{5.3}$$

For example, if $M_r = 100$, $\tau_c \approx 100$ ps; if $M_r = 10\,000$, $\tau_c \approx 10$ ns. On a 400 MHz ($\omega_0/2\pi = 4 \times 10^8\ \text{s}^{-1}$) spectrometer, the maximum relaxation rate occurs when $\tau_c \approx 400$ ps, so that most molecules with an M_r of a couple of hundred or less should fall to the left of the T_1 minimum, while those with a molecular weight of a thousand or more come on the right.

It can now be seen why *rotational* motion is important for most spin relaxation mechanisms. *Vibrations* are usually much too fast to have a significant component at the relatively low frequencies involved in NMR. Modulation of *inter*molecular dipolar interactions by *translational* motion is also relatively inefficient because the couplings are generally weaker

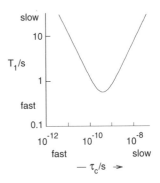

Fig. 5.5 The dependence of the spin–lattice relaxation time T_1 on the rotational correlation time τ_c, from eqns 5.2 and 5.4, using $\gamma^2\langle B^2\rangle = 4.5 \times 10^9\ \text{s}^{-2}$ and $\omega_0/2\pi = 400$ MHz. Both axes are logarithmic. The regions of the graph corresponding to fast and slow tumbling, and fast and slow relaxation are indicated. The value of $\gamma^2\langle B^2\rangle$ used is roughly appropriate for the dipolar coupling of two protons separated by 2 Å.

(larger average nuclear separation) than in the intramolecular case. *Rotation*, by contrast, occurs at roughly the right frequency, modulates *intramolecular* interactions, and so is near optimal for relaxation.

To see how spin–lattice relaxation depends on the local magnetic fields that cause it, we look at the predicted relaxation rate for the idealized random fields mechanism (Slichter 1990):

$$\frac{1}{T_1} = \gamma^2 \langle B^2 \rangle \, J(\omega_0) \tag{5.4}$$

where the spectral density $J(\omega_0)$ has been discussed in the preceding section and $\langle B^2 \rangle$ is the mean square value of the local fluctuating fields. Not surprisingly, stronger local fields lead to faster relaxation, other things being equal.

Using eqn 5.4 and the form of the magnetic dipolar coupling, eqn 3.11, it is clear that the spin–lattice relaxation rate of a pair of interacting nuclei should be proportional to r^{-6} (i.e. r^{-3} squared). T_1s are thus sensitive functions of internuclear separations and hence molecular structure. Further, the mean square dipolar field produced by a spin S is proportional to γ_S^2 (eqns 3.11 and 1.5), so that the spin–lattice relaxation rate of a nearby spin I should be proportional to $\gamma_I^2 \gamma_S^2$ (using eqn 5.4). A proton therefore relaxes a nearby ^{13}C much less efficiently than a ^1H at the same distance.

The predictions of this rough and ready argument are confirmed by a more detailed theoretical treatment which gives the following expression for the spin–lattice relaxation rate of two dipolar-coupled nuclei, I and S, with different gyromagnetic ratios, in the so-called *extreme narrowing limit* $\omega_0 \tau_c \ll 1$

$$\frac{1}{T_1} = \left(\frac{\mu_0}{4\pi} \right)^2 \frac{\gamma_I^2 \gamma_S^2 \hbar^2 \tau_c}{r^6} \tag{5.5}$$

The τ_c part of eqn 5.5 comes from the spectral density (eqns 5.2 and 5.4). Typical dipolar T_1 values for protons are in the region of 0.1–10 s (see also Fig. 5.5).

Nuclear spin relaxation is slow for two reasons. First, the local magnetic fields are generally rather feeble, and interact weakly with nuclear spins. Second, the interactions must be rapidly modulated to be effective; and even when $\omega_0 \tau_c = 1$, the spectral density at the NMR frequency is small.

There are two further points about spin–lattice relaxation. First, when a collection of nuclei relaxes, energy is either released or absorbed by the spin system as the populations return to their equilibrium values. Where does this energy go to or come from? Second, why do the transitions induced by local time-dependent fields return the spin system to equilibrium?

Spin–lattice relaxation couples the spins (very weakly) to the motion of the molecules that carry them and so provides a pathway for the exchange of energy between the spin system and its surroundings. In other words, spin–lattice relaxation brings the spins into thermal contact with the lattice, enabling them to come to equilibrium with the rest of the world. The energy absorbed or released in the course of spin relaxation is transferred from or to the motions the molecules, causing a slight cooling or warming of the lattice. Since the spin energies are minute compared to the rotational, vibrational and translational energy of molecules in solution, nuclear spins are relaxed with an unmeasurably small change in the temperature of the sample, just as dropping a small hot object into a large lake would cause a negligible rise in the temperature of the water.

Some illustrations and applications of spin–lattice relaxation are presented in Sections 5.7 and 5.8; but first we look at an important relaxation phenomenon resulting from the dipolar relaxation mechanism.

5.4 The nuclear Overhauser effect

Consider a molecule containing two inequivalent protons, I and S, with no *scalar* coupling, so that the ^{1}H spectrum consists of a singlet at each of the chemical shifts. Suppose that, while recording the spectrum, the S spins are saturated (i.e. their $m = +\frac{1}{2}$ and $m = -\frac{1}{2}$ populations are equalized) by applying a strong radiofrequency field oscillating at the S resonance frequency. Of course, this destroys the NMR signal of S, but it can also affect the intensity of the I resonance if the two spins have an appreciable *dipolar* interaction. As shown schematically in Fig. 5.6, the I peak may get stronger, or weaker, or even invert. This remarkable phenomenon is known as the *nuclear Overhauser effect* (or *enhancement*), NOE. As we shall see, it gives information on internuclear separations much more directly than spin–lattice relaxation.

To understand the origin of the NOE, it is necessary to look at the possible spin–lattice relaxation pathways available to a pair of dipolar coupled spin-$\frac{1}{2}$ nuclei, I and S. Figure 5.7 shows the four energy levels: $\alpha_I\alpha_S$, $\alpha_I\beta_S$, $\beta_I\alpha_S$, $\beta_I\beta_S$. Ignoring the small chemical shift difference, the $\alpha_I\alpha_S$ and $\beta_I\beta_S$ states have energies $-\omega_0$ and $+\omega_0$ respectively (in angular frequency units), while $\alpha_I\beta_S$ and $\beta_I\alpha_S$ are midway between, with zero energy. At thermal equilibrium, the relative populations of the four states, to a very good approximation (eqn 1.12), are $1+2\Delta$ ($\alpha_I\alpha_S$); 1 ($\alpha_I\beta_S$ and $\beta_I\alpha_S$); $1-2\Delta$ ($\beta_I\beta_S$), where $\Delta = \frac{1}{2}\hbar\omega_0/kT$.

Of the six relaxation pathways open to the two coupled spins, four correspond to a *single* spin flipping, i.e. $\alpha_I \leftrightarrow \beta_I$ or $\alpha_S \leftrightarrow \beta_S$, and are nothing more than the spin–lattice relaxation processes discussed above. Their rates are denoted W_1^I and W_1^S, where the subscript indicates that the magnetic quantum number changes by ± 1, and the superscript identifies the spin that flips.

The other two relaxation pathways are *cross relaxation* processes in which I and S relax *together*, i.e. $\alpha_I\alpha_S \leftrightarrow \beta_I\beta_S$ (both spins flipping in the same direction at rate W_2^{IS}) and $\alpha_I\beta_S \leftrightarrow \beta_I\alpha_S$ (I and S flipping in opposite directions, at rate W_0^{IS}). Cross relaxation comes about because the chaotic molecular motion, combined with the mutual dipolar interaction, causes the fluctuating local fields experienced by I and S to be *correlated*— $(3\cos^2\theta - 1)/r^3$ is the same for both spins—with the result that the nuclei can undergo *simultaneous* spin–flips. The W_0^{IS} and W_2^{IS} processes are simply

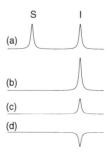

Fig. 5.6 Schematic NMR spectra showing various possible nuclear Overhauser enhancements. (a) Conventional spectrum of two neighbouring spins I and S. (b)–(d) Possible spectra resulting from saturation of the S resonance: the I peak either gets stronger (b), weaker (c), or inverts (d), depending on the conditions.

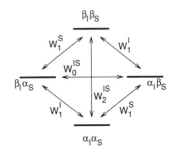

Fig. 5.7 Energy levels for a pair of spin-$\frac{1}{2}$ nuclei I and S, showing the six possible relaxation pathways.

Fig. 5.8 Spin state populations for a pair of neighbouring spin-$\frac{1}{2}$ nuclei I and S. Shaded circles indicate a population excess of $\Delta = \frac{1}{2}\hbar\omega_0/kT$; open circles, a population deficit of Δ. The population differences for the I spin transitions are shown next to the arrows. (a) Thermal equilibrium. (b) Effect of saturating both transitions of spin S. (c) Effect of rapid $\alpha_I\alpha_S \leftrightarrow \beta_I\beta_S$ cross relaxation. (d) Effect of rapid $\alpha_I\beta_S \leftrightarrow \beta_I\alpha_S$ cross relaxation.

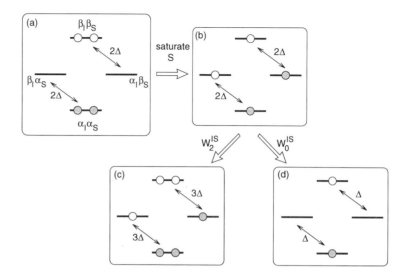

extra pathways that allow the spin state populations to return to equilibrium following some disturbance. Note that spin relaxation does not involve absorption or emission of electromagnetic radiation, and so is not subject to the usual $\Delta m = \pm 1$ selection rule.

Now, to see how the NOE arises, imagine applying a radiofrequency field to spin S of sufficient strength to saturate both S transitions ($\alpha_I\alpha_S \leftrightarrow \alpha_I\beta_S$ and $\beta_I\alpha_S \leftrightarrow \beta_I\beta_S$), i.e. to equalize the populations of $\alpha_I\alpha_S$ and $\alpha_I\beta_S$, and of $\beta_I\alpha_S$ and $\beta_I\beta_S$, Fig. 5.8(a) and (b). This has no effect on the population differences across the I transitions ($\alpha_I\alpha_S \leftrightarrow \beta_I\alpha_S$ and $\alpha_I\beta_S \leftrightarrow \beta_I\beta_S$), which are still 2Δ. In Fig. 5.8, a solid circle is used to denote a population *excess* of Δ, while an open circle indicated a population *deficit* of the same size. Now suppose that all relaxation pathways are insignificant except W_2^{IS} (an unrealistic but convenient approximation). This relaxation route transfers population between $\alpha_I\alpha_S$ and $\beta_I\beta_S$ and restores the equilibrium populations of those two states, $1+2\Delta$ and $1-2\Delta$ respectively, Fig. 5.8(c). The population differences across the I transitions are now 3Δ; that is, the intensity of the I peak has *increased* by 50%. Cross relaxation has transferred magnetization from the saturated spin S to its dipolar-coupled partner I.

Conversely, if S is saturated and the W_0^{IS} process is dominant, the populations of $\alpha_I\beta_S$ and $\beta_I\alpha_S$ are restored to their equilibrium values (both unity) giving a population difference of only Δ across the I transitions, i.e. a 50% *reduction* in the intensity of the I resonance, Fig. 5.8(d).

The NOE can be quantified by a parameter η, defined in terms of i the perturbed NMR intensity of spin I, and i_0 its normal intensity:

$$\eta = \frac{i - i_0}{i_0}. \tag{5.6}$$

The simplified and approximate argument above gives extreme values for η of $+\frac{1}{2}$ and $-\frac{1}{2}$. An exact treatment shows that the maximum homonuclear

NOE is indeed $+\frac{1}{2}$, but that the minimum is -1. In reality, neither W_2^{IS} nor W_0^{IS} dominates the other relaxation pathways and η is somewhere between the extremes. It can be seen from the above discussion that η has the same sign as $W_2^{IS} - W_0^{IS}$.

But what determines the size and sign of $W_2^{IS} - W_0^{IS}$? We saw in Section 5.3 that the rate of spin–lattice relaxation is proportional to the spectral density at the frequency of the transition. So

$$W_2^{IS} \propto J(2\omega_0) \quad \text{and} \quad W_0^{IS} \propto J(0). \tag{5.7}$$

Consider first a large molecule with an average tumbling rate τ_c^{-1} much less than ω_0. As indicated in Fig. 5.9(a), the spectral density at $2\omega_0$ and therefore W_2^{IS}, are negligible, so that $W_2^{IS} - W_0^{IS} < 0$, leading to a negative enhancement η. In other words, saturating a proton in a large molecule should cause a *reduction* in the NMR intensity of nearby protons.

For a small molecule, the situation is not quite so clear. If the average tumbling rate τ_c^{-1} is much greater than ω_0, then there is essentially an equal chance of finding the molecule tumbling at the two relevant frequencies, zero and $2\omega_0$ (Fig. 5.9(a)). Thus, in the extreme narrowing limit, the difference between the two cross relaxation rates arises not from spectral densities but from the ability of the dipolar interaction to induce the two relaxation processes, $\alpha_I\alpha_S \leftrightarrow \beta_I\beta_S$ and $\alpha_I\beta_S \leftrightarrow \beta_I\alpha_S$. It turns out that when $\omega_0\tau_c \ll 1$, W_2^{IS} is *six times larger* than W_0^{IS}, giving a positive enhancement, η (Fig. 5.9(b)). Thus, saturating a proton in a small molecule should boost the NMR intensity of neighbouring protons. Infuriatingly, there is no satisfactory handwaving explanation for the origin of this factor of six.

To summarize, the proton–proton NOE, η, should be positive for fast motion and negative for slow motion. The change of sign occurs when $W_0^{IS} = W_2^{IS}$, at which point the effects of the two cross relaxation pathways cancel; this happens when $\omega_0\tau_c \approx 1$ (Fig. 5.10).

So far we have restricted attention to the dipolar relaxation mechanism for the very good reason that it is by far the most common source of cross relaxation. In fact, most other relaxation mechanisms tend to diminish NOEs by returning population differences to equilibrium via the W_1 pathways, so short-circuiting the cross relaxation routes that are crucial for the NOE. The extreme values of the homonuclear NOE in the limits of very fast and very slow motion are therefore only to be expected when the dipolar interaction is the dominant source of relaxation.

NOEs are also observable for heteronuclear pairs of spins; for example, saturation of proton resonances produces a welcome enhancement in the NMR signals of nearby ^{13}C nuclei. In the extreme narrowing limit, the enhancement, in the absence of other relaxation mechanisms, is

$$\eta = \frac{1}{2}\frac{\gamma_S}{\gamma_I} \tag{5.8}$$

where γ_I and γ_S are the gyromagnetic ratios, S is the saturated nucleus and I the observed nucleus. Thus for $I = {}^{13}C$ and $S = {}^1H$, $\gamma_H/\gamma_C \approx 4$ and the maximum NOE is roughly 2; in the homonuclear case $\eta = \frac{1}{2}$, as described above.

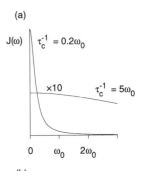

(a)

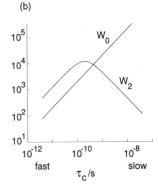

(b)

Fig. 5.9 (a) Spectral densities $J(\omega)$ as a function of ω for $\tau_c^{-1} = 0.2\omega_0$ (slow tumbling), and $\tau_c^{-1} = 5\omega_0$ (fast tumbling). Note that the latter is plotted on an expanded vertical scale. (b) Dependence of the two cross relaxation rates W_0^{IS} and W_2^{IS} on τ_c, calculated for $\omega_0/2\pi = 400$ MHz. Both axes are logarithmic. The units for the vertical axis are arbitrary. The maximum in W_2 occurs at $\tau_c = (2\omega_0)^{-1}$.

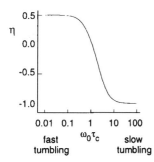

Fig. 5.10 Dependence of the nuclear Overhauser enhancement η (eqn 5.6) on $\omega_0\tau_c$. This is the maximum homonuclear NOE, observable only in the absence of relaxation mechanisms other than the dipolar mechanism. Note that the horizontal axis is logarithmic, and that η = 0 when $\omega_0\tau_c \approx 1$.

As Sections 5.7 and 5.8 will reveal, the NOE is exceedingly useful as a source of information on molecular structure. Although the r^{-6} dependence of the cross relaxation rates (by analogy with eqn 5.5) in principle gives internuclear distances directly, in practice matters are a little more involved. Estimates of *relative* separations can often be obtained if the pairs of nuclei concerned are undergoing similar motions *and* if they have comparable contributions from other, non-dipolar, relaxation mechanisms. NOEs are also often used semiquantitatively to decide whether two nuclei are close to one another or not. Despite these limitations, the NOE is still an invaluable source of structural data.

Finally it should be said that selective saturation is certainly not the only way of measuring NOEs. A more satisfactory way of disturbing spin populations and observing magnetization transfer to nearby nuclei is outlined in Section 6.4. For more on the NOE, see Neuhaus and Williamson (1989).

5.5 Spin–spin relaxation

In Chapter 4, we saw that slow chemical exchange reduces the lifetime of spin states and causes linebroadening. Clearly spin–lattice relaxation should have the same effect, and a broadening of the order of $1/\pi T_1$ (see eqn 4.2) can be expected. However, this is not the only way relaxation processes affect NMR linewidths, and so it is useful to define a new parameter, the *spin–spin relaxation time*, T_2

$$\frac{1}{\pi T_2} = \Delta\nu \qquad (5.9)$$

where $\Delta\nu$ is the linewidth associated with relaxation processes (instrumental contributions are mentioned in Section 6.1). T_2 has a more fundamental interpretation, but for now we regard it as a linewidth parameter.

The second contribution to the linewidth, and hence T_2, may be understood from the following argument. Imagine a disordered molecular solid in which all motions are frozen out. Every nucleus has several neighbours, to each of which it has a dipolar coupling proportional to $(3\cos^2\theta - 1)/r^3$ (Section 3.8). The NMR line of every spin is therefore split many times over by the dipolar interactions with neighbouring spins each with different r and θ. When this already complicated pattern is summed over all possible orientations of the molecules in the disordered solid, the result is a single, largely featureless peak with a width related to the root-mean-square dipolar interaction, which might be several kilohertz. From eqn 5.9, T_2 would be rather small, perhaps a millisecond or less.

Now, suppose the solid is allowed to melt. As the molecules tumble around $(3\cos^2\theta - 1)/r^3$, and hence the dipolar coupling of every pair of nuclei, is rapidly modulated. By analogy with fast chemical exchange (Section 4.1), this should average the dipolar splittings over all possible values of r and θ, *provided* the motion is fast compared to the frequency differences being modulated, i.e. the dipolar linewidth discussed above. Since the spherical average of $(3\cos^2\theta - 1)$ is zero (eqn 3.14), it follows that the molecular motion should reduce the dipolar broadening, i.e. T_2 (eqn 5.9) should *increase* as the rotational correlation time τ_c *decreases*.

One can even use the theory of chemical exchange to estimate the size of this contribution to T_2. According to eqn 4.3, the linewidth in fast exchange is roughly $(\delta v)^2 \tau$ where δv is the difference in resonance frequencies and τ is the mean lifetime at each frequency. By analogy, the residual dipolar linewidth for a molecule in solution should be roughly $\gamma^2 \langle B^2 \rangle \tau_c$, i.e. the mean square dipolar splitting multiplied by the rotational correlation time of the motion.

For the random fields relaxation mechanism, theory (Slichter 1990) gives

$$\frac{1}{T_2} = \tfrac{1}{2}\gamma^2 \langle B^2 \rangle \, J(\omega_0) + \tfrac{1}{2}\gamma^2 \langle B^2 \rangle \, J(0) \ . \tag{5.10}$$

The first part of this expression is simply $\tfrac{1}{2}T_1^{-1}$ (eqn 5.4) and represents the lifetime broadening caused by spin–lattice relaxation. The second term is the contribution discussed above and is indeed equal to $\gamma^2 \langle B^2 \rangle \tau_c$ if, as in eqn 5.2, $J(0) = 2\tau_c$.

The motional dependence of T_2 is shown in Fig. 5.11, together with the corresponding T_1 behaviour from eqn 5.4. As anticipated, T_2 increases as the tumbling gets faster and more effectively averages the residual dipolar broadening. In the slow motion limit ($\omega_0 \tau_c \gg 1$) the component of T_2 arising from spin–lattice relaxation is negligible (because $J(\omega_0) \ll J(0)$, see Fig. 5.4) and the spin–spin relaxation rate is simply proportional to the correlation time ($J(0) = 2\tau_c$).

Interestingly, the two relaxation times are identical in the extreme narrowing limit, when the tumbling is fast compared to the resonance frequency ω_0. When $\omega_0 \tau_c \ll 1$, $J(\omega_0)$ and $J(0)$ are equal and so, therefore, are the two terms in eqn 5.10. Comparing this with eqn 5.4, it is clear that $T_1 = T_2$. Irritatingly, there seems to be no satisfactory physical reason why the second term in eqn 5.10 should equal $\tfrac{1}{2}T_1^{-1}$ in the limit of very fast tumbling.

Although almost everything said hitherto has been about the dipolar mechanism, there are other sources of relaxation. Any magnetic interaction experienced by nuclear spins can, in principle, lead to relaxation provided random molecular motions are able to produce an appropriate time dependence. Examples include modulation of the anisotropic chemical shift interaction by molecular tumbling; modulation of scalar couplings by internal motion; and interactions with the strong fields generated by unpaired electrons in paramagnetic species. For $I > \tfrac{1}{2}$ nuclei, however, there is an additional, often dominant, relaxation mechanism to which we now turn.

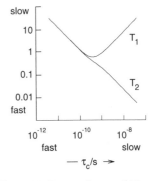

Fig. 5.11 Dependence of T_1 and T_2 on rotational correlation time τ_c, from (eqns 5.4 and 5.10) using $\gamma^2 \langle B^2 \rangle = 4.5 \times 10^9 \text{ s}^{-2}$ and $\omega_0/2\pi = 400$ MHz. Both axes are logarithmic. The units for the vertical axis are seconds. The regions of the graph corresponding to fast and slow tumbling, and fast and slow relaxation are indicated.

5.6 Quadrupolar relaxation

A nucleus with a spin quantum number greater than $\tfrac{1}{2}$ possesses an *electric quadrupole moment* in addition to its magnetic dipole moment. One can

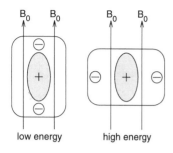

Fig. 5.12 A quadrupolar nucleus, whose non-uniform charge distribution is represented by an elipsoid, in the electric field of two negative charges in a molecule, shown as a box. As the molecule rotates, the spin axis remains aligned along the B_0 field, and the energy of the nucleus is modulated as indicated.

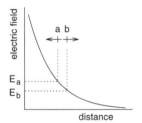

Fig. 5.13 The interaction, with a non-uniform electric field, of an electric quadrupole, regarded as two barely separated back-to-back dipoles a and b. The net interaction is proportional to the *difference* in electric fields experienced by the two dipoles $(E_a - E_b)$ which, in turn, is determined by the *gradient* of the electric field at the position of the quadrupole.

think of this in terms of an elipsoidal charge distribution in the nucleus (Fig. 5.12) with an excess of positive charge near the north and south poles and a corresponding depletion around the equator. Unlike electric dipoles—for example polar molecules like HCl—electric quadrupoles do not interact with spatially uniform electric fields, but only with *electric field gradients*, a property that may be understood by regarding the nuclear quadrupole moment as two identical back-to-back electric dipoles (Fig. 5.13).

The electric field gradient at a nucleus is a measure of the *non-uniform* distribution of local electronic charge: in sufficiently high symmetry environments—spherical, cubic, octahedral or tetrahedral—the electric field gradients generated by surrounding charges exactly cancel out, giving no net quadrupolar interaction. Nuclei in lower symmetry environments, however, experience non-zero electric field gradients which depend on the orientation of the molecule in the magnetic field of the NMR spectrometer. For example (Fig. 5.12), two negative charges at a fixed distance from a quadrupolar nucleus have a more favourable Coulombic interaction when they lie on the spin axis, i.e. close to the poles of the non-spherical nucleus, than when they are in the equatorial plane. This anisotropic interaction, like the magnetic dipolar coupling discussed in Section 3.8, produces splittings in the NMR spectra of single crystals, and gives broad lines for powders and disordered solids. Quadrupolar interactions also cause *spin relaxation* when modulated by molecular tumbling.

Many $I > \frac{1}{2}$ nuclei in low symmetry environments have large quadrupolar interactions and consequently very efficient spin relaxation. The most obvious manifestations of this are the large linewidths (small T_2) generally observed for nuclei such as ^{14}N, ^{17}O, ^{35}Cl, ^{37}Cl. For example, the ^{14}N resonance of the pyramidal NMe_3 is almost 100 Hz wide, while $^{14}NMe_4^+$, $^{15}NMe_3$, and $^{15}NMe_4^+$ all have nitrogen NMR linewidths less than 1 Hz ($^{14}NMe_4^+$ is tetrahedral and has no electric field gradient; ^{15}N has $I = \frac{1}{2}$ and therefore no quadrupole moment). Such linebroadenings severely reduce resolution, and account for the relative unpopularity of quadrupolar nuclei in NMR studies of liquids.

The other major consequence of quadrupolar relaxation is the loss of multiplet structure for spins that are scalar-coupled to quadrupolar nuclei. Efficient spin–lattice relaxation of, for example, an $I = 1$ nucleus causes it to flip rapidly between its three spin orientations ($m = +1, 0, -1$) so that a coupled spin (chemical shift frequency ν_0, coupling constant J) has a resonance frequency that jumps rapidly between $\nu_0 + J$, ν_0, and $\nu_0 - J$. If this happens at a rate fast compared to the frequency differences involved, i.e. J and $2J$, then one should see just a *single line* at the mean frequency, ν_0 (see the discussion of chemical exchange effects in Chapter 4). This is essentially identical to the collapse of multiplet splittings brought about by fast intermolecular proton exchange in the ^1H spectrum of ethanol (Section 4.3 and Fig. 4.12). Examples of molecules in which quadrupolar nuclei do not lead to multiplet splittings can be found in Fig. 2.1 (^{35}Cl and ^{37}Cl), Fig. 3.16 (^{14}N, ^{79}Br, and ^{81}Br), and Fig. 4.10 (^{99}Ru and ^{101}Ru).

5.7 Examples—structure

The applications of relaxation effects in NMR usually fall into one of two categories—structure and dynamics. Roughly speaking, T_1 measurements are more useful for motional studies, and NOEs for structural investigations, although there is plenty of overlap. In general, to get information on the structure and/or conformation of a molecule, one needs to have some knowledge of, or make assumptions about, its dynamical properties, and vice versa. For example, it is very difficult to use NOEs to determine the average conformations of highly flexible molecules.

Quaternary carbons

The ^{13}C spin–lattice relaxation times of organic molecules are often dominated by ^{13}C–^1H dipolar interactions, the closest protons having the greatest effect because of the r_{CH}^{-6} distance dependence (eqn 5.5). Quaternary carbons, which have no directly bonded protons, therefore usually relax more slowly than CH and CH_2 carbons in the same molecule, at least in the absence of appreciable internal motion. For example, the four types of quaternary carbons in fluoranthene (Fig. 5.14) relax five to eight times slower than the five methines.

A similar picture emerges from the $^{13}C\{^1H\}$ NOEs (Fig. 5.14). The enhancements for the methine carbons are all close to the maximum value of 1.99 as expected for a predominantly dipolar relaxation mechanism. The smaller NOEs, for the quaternary carbons, are consistent with slower dipolar relaxation and a relatively larger contribution from other mechanisms which do not cause cross relaxation. Carbons *f*, *g* and *i* are enhanced by protons on adjacent carbons (*d*, *a* and *c* respectively), while *h* must rely on even more distant protons, hence its small NOE.

CH and CH_2 carbons

In a rigid molecule, a methylene carbon with dipolar interactions to two directly bonded protons should relax *twice* as fast as a methine carbon bearing a single proton, other things being equal. For example, the ratio of ^{13}CH to $^{13}CH_2$ spin–lattice relaxation times in adamantane (Fig. 5.15) is indeed close to 2.0. The actual ratio of 1.8 is accounted for by considering the different number of β protons (on adjacent carbons) in the two cases: six for the CHs and two for the CH_2s. The extra dipolar relaxation due to the β protons is thus three times more effective for the CH carbons, causing then to relax more rapidly than would be expected simply on the basis of the α protons. A simple calculation using standard bond lengths and angles shows that the ratio of T_1s (CH : CH_2) should be reduced from 2.0 to 1.82 in excellent agreement with the experimental value. Finally it must be said that ^{13}C spin–lattice relaxation is not always as straightforward as it is in adamantane. In more complex molecules of lower symmetry, anisotropic tumbling and internal motions (see below) often serve to complicate matters.

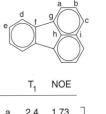

	T_1	NOE	
a	2.4	1.73	methine
b	2.4	1.98	
c	2.1	1.86	
d	2.5	1.72	
e	2.2	1.86	
f	16.1	0.33	quaternary
g	15.6	0.34	
h	15.1	0.15	
i	11.6	0.33	

Fig. 5.14 ^{13}C spin–lattice relaxation times (seconds) and $^{13}C\{^1H\}$ nuclear Overhauser enhancements for fluoranthene. (Data from C. Yu and G. C. Levy, *J. Amer. Chem. Soc.*, 1984, **106**, 6533.)

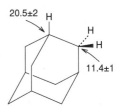

Fig. 5.15 ^{13}C spin–lattice relaxation times (seconds) of the CH and CH_2 carbons of adamantane. (Data from K. F. Kuhlmann, D. M. Grant, and R. K. Harris, *J. Chem. Phys.*, 1970, **52**, 3439.)

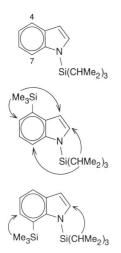

Fig. 5.16 Two possible products of trimethylsilylation of *N*-triisopropylsilylindole (top) showing the nuclear Overhauser enhancements that would result from saturation of the SiMe$_3$ methyl protons and the Si–Pr$^i_3$ methine proton. (See G. Nechvatal and D. A. Widdowson, *J. Chem. Commun.*, 1982, 467; and D. Neuhaus and M. P. Williamson, *The Nuclear Overhauser Effect*, VCH Publishers, New York, 1989, pp 362–3.)

The r^{-6} distance dependence of the dipolar relaxation mechanism is an invaluable source of information on the structures and conformations of molecules. By far the most useful relaxation parameter is the NOE, which under the right conditions, is governed by a single relaxation mechanism (dipolar) and by a single internuclear distance. Spin–lattice relaxation times are less informative, being determined by the dipolar interactions to all nearby nuclei, as well as by other relaxation mechanisms.

Determining structure from NOEs

A difficulty routinely faced by chemists is to decide whether a newly synthesized molecule is the desired compound or a closely related one. In many cases, the structure will be known except for some small but crucial detail like the stereochemistry or the positioning of a substituent. In many cases, the NOE provides a quick and unambiguous answer.

Trimethylsilylation of the triisopropylsilylindole in Fig. 5.16 can lead to two possible trimethylsilyl derivatives, with either 4 or 7 substitution. Both isomers have five inequivalent aromatic protons, and therefore four doublets and a doublet of doublets in the ^{1}H spectrum. Unless one were happy to rely on subtle chemical shift arguments, it would be difficult to decide which compound had been formed.

The solution is provided by two quick NOE experiments: irradiation of the trimethylsilyl singlet boosts the intensity of two of the four doublets, while irradiation of the septet from the CH proton in the triisopropylsilyl group enhances the *other* two doublets. These observations are only consistent with 4-substitution. The 7-trimethylsilyl derivative would have shown enhancements for just one doublet in each experiment.

Protein structure determination

One of the most impressive achievements of NMR in recent years has been the determination of the three-dimensional structure of proteins in solution. Starting from the sequence of amino acids and a set of NMR spectra it is now possible to define the conformations of proteins of up to about 20 000 molecular weight, containing more than 150 amino acid residues.

The trick is to detect many hundreds of NOEs between pairs of protons at known positions in the molecule. Some of these will link nuclei in the same residue, or in neighbouring residues in the sequence, but many will connect protons in very different parts of the molecule. These NOE 'connectivities' allow one to set upper limits on the separations of the protons involved, which may vary from 2 to 5 Å, depending on the circumstances. Even though these distance constraints are short range compared to the overall dimensions of the protein (typically a few tens of angstroms) and often rather imprecise, if there are enough of them distributed throughout the protein they define the structure very effectively (Fig. 5.17). All that is required is a sophisticated computer algorithm to search for the conformations that satisfy all the NOE constraints, together with any restrictions on torsion angles from three-bond coupling constants.

The whole procedure relies on having first *assigned* a substantial fraction of the ¹H spectrum: i.e. as many resonances as possible must be resolved and attributed to specific protons in the sequence. The initial stage of this formidable task is to identify NMR lines corresponding to protons in the *same* amino acid residue. This is done by mapping out the network of scalar couplings in each residue, knowing that pairs of protons only have appreciable spin–spin coupling constants if they are separated by two or three chemical bonds. For example, alanines are easily identified because Ala is the only amino acid whose αCH proton and side-chain (a methyl group) make up an AX_3 spin system.

The next stage is to link together the individual amino acid spin systems by detecting NOEs between protons in consecutive residues, e.g. between the amide NH protons, which are separated by 2–5 Å. In this way the spectrum can be assigned sequentially, stepping from one residue to the next.

Two sorts of NMR experiment are therefore required: one to determine the *through-bond* connectivities (*J*-couplings), and a second to establish *through-space* connectivities (NOEs). Although this sort of information can be obtained for small molecules by a series of experiments with selective excitation, for example saturating in turn each resonance and looking for changes elsewhere in the spectrum, this is not feasible for proteins which may have more than 1000 NMR lines crammed into 10 ppm, with substantial resonance overlap. In practice, protein structure determination has been made possible by the development of routine *two-dimensional* NMR techniques, in which resonances are spread out into a second frequency dimension, to alleviate overcrowding and to allow all connectivities of a particular type (*J*-couplings or NOEs), to be obtained in a single experiment without the need for frequency-selective irradiation. The basis of these revolutionary techniques is outlined in Section 6.4.

5.8 Examples—dynamics

As mentioned above, information on molecular motions can usually only be extracted from relaxation data when the structure is well defined. Many relaxation studies therefore use 'probes' of known geometry, with a dominant (usually dipolar) relaxation mechanism, for example ¹³C in CH groups, ¹⁵N in NH groups, etc.

Methyl group internal rotation

In one of the examples above, we saw that CH_2 carbons often relax about twice as rapidly as CH carbons. Extrapolating, we might expect that a methyl carbon which is relaxed by dipolar couplings to three directly bonded protons might relax three times as fast as a ¹³CH group in the same molecule. However this is rarely the case because of the rapid internal rotation of methyl groups. For example, the dipolar relaxation of the methyl carbons in mesitylene (Fig. 5.18) relax *more slowly* than do the ring carbons. This is easily understood when one remembers that for extreme

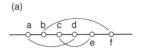

(a)

(b)

Fig. 5.17 Protein structure determination from nuclear Overhauser effects (schematic). NOEs are detected between nuclei at different positions along the polypeptide chain (a), which must therefore be close to one another. A computer search algorithm finds structures which satisfy all these distance constraints (b).

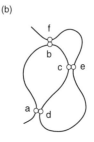

Fig. 5.18 ¹³C spin–lattice relaxation times of CH and CH_3 carbons in mesitylene (top) and *ortho*-xylene (bottom). Only the dipolar contributions to the relaxation times are shown. (Data from K. F. Kuhlmann and D. M. Grant, *J. Chem. Phys.*, 1971, **55**, 2998.)

narrowing, faster motion means slower relaxation. The ring carbons which enjoy no internal motion must rely on the relatively slow overall tumbling of the molecule to modulate the $^{13}C-^{1}H$ dipole interactions and cause relaxation. The methyl groups, however, rotate much faster and give more effective averaging of the dipolar interactions and hence slower relaxation.

In *ortho*-xylene (Fig. 5.18), however, the methyl and methine ^{13}C T_1s are very similar because steric interactions between the adjacent methyl groups slow down the internal motion and accelerate methyl carbon relaxation relative to the ring carbons. Analysis of the data for *ortho*-xylene shows that the internal rotation is roughly twelve times faster than overall tumbling and has a barrier height of about 6 kJ mol^{-1}.

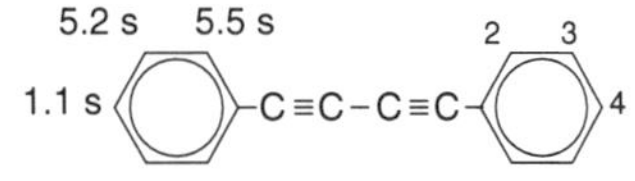

Anisotropic rotation

Figure 5.19 shows the ^{13}C spin–lattice relaxation times for the phenyl carbons in diphenyldiacetylene. From the foregoing discussion one might anticipate that the carbons in position 2 (C2) might relax a little more slowly than C3 and C4 because of the different number of β protons. In fact C2 and C3 have very similar T_1s, which are almost five times longer than C4.

The origin of this effect lies in the anisotropic tumbling of this rod-like molecule: the end-over-end motion is much slower than rotation around the long axis. The relaxation of C4 is predominantly caused by motions that modulate the dipolar coupling to its attached proton, i.e. by motions that change the angle between this CH bond and the magnetic field direction. Clearly the rapid long-axis rotation is ineffective in this respect, so that it is the end-over-end motion that relaxes C4. Since this motion is slow, and the molecule falls in the extreme narrowing limit, the relaxation of C4 is fast. By contrast, the CH vectors of C2 and C3 point away from the long axis, allowing the rapid motion around this axis to modulate the CH dipolar coupling, resulting in slow relaxation. The end-over-end motion is about twenty times slower than that around the long axis.

Of course this is a fairly extreme example of anisotropic motion: not all molecules deviate as much from spherical symmetry as this one. However the effects of anisotropic motions can be discerned in many small molecules.

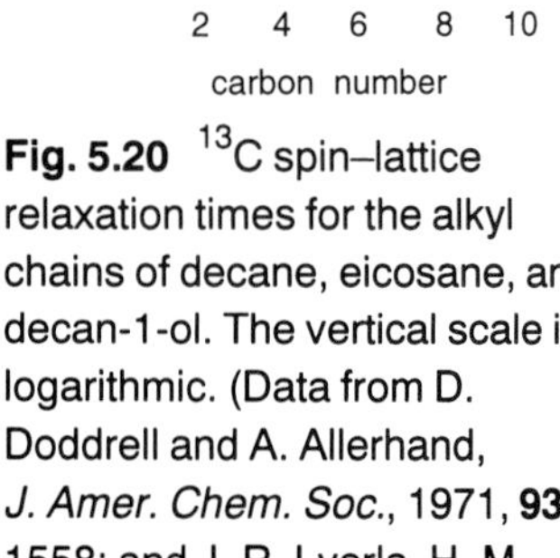

5.2 s 5.5 s

1.1 s

2 3 4

Fig. 5.19 ^{13}C spin–lattice relaxation times of the C2, C3, and C4 ring carbons in diphenyldiacetylene. (Data from G. C. Levy, J. D. Cargioli, and F. A. L. Anet, *J. Amer. Chem. Soc.* 1973, **95**, 1527.)

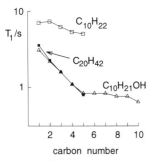

Fig. 5.20 ^{13}C spin–lattice relaxation times for the alkyl chains of decane, eicosane, and decan-1-ol. The vertical scale is logarithmic. (Data from D. Doddrell and A. Allerhand, *J. Amer. Chem. Soc.*, 1971, **93**, 1558; and J. R. Lyerla, H. M. McIntyre, and D. A. Torchia, *Macromolecules*, 1974, **7**, 11.)

Alkyl chains

^{13}C spin–lattice relaxation can be used to study the motions of highly flexible molecules. For example, Fig. 5.20 shows T_1s for the carbons in the linear compounds decane ($C_{10}H_{22}$), eicosane ($C_{20}H_{42}$), and decan-1-ol ($C_{10}H_{21}OH$). These data can be understood in terms of the *overall motion* of the molecule, which becomes slower as the molecular weight and viscosity increase, and the *internal motion* due to rotation about individual C–C bonds, which is fastest at the chain ends.

Comparing the two alkanes, eicosane tumbles slowly, hence the generally short T_1s, and has relatively rapid internal motion which causes a strong variation in T_1 along the chain, with the slowest relaxation at the mobile chain ends. In decane, which tumbles more rapidly, the two types of motion occur at more similar rates, leading to a reduced importance of internal motion and a smaller spread in T_1s.

In decanol, the T_1s are all shorter than for decane because of the increased viscosity, and decrease steadily towards the OH group, reflecting the restriction of internal motion caused by hydrogen bonding.

Solid benzene

The ^{1}H NMR spectrum of solid benzene below 90 K consists of a single line some 40 kHz wide, the result of a multitude of strong unresolved intra- and intermolecular dipolar splittings. As shown in Fig. 5.21(a), the line sharpens dramatically between 90 and 120 K, and then remains more or less unchanged up to the melting point, where rapid tumbling reduces the width to less than 1 Hz (see Section 5.5). The linewidth change in the solid is due to molecular reorientation around the six-fold axis of the molecule which partially averages the dipolar interactions. Narrowing occurs when the frequency of rotation becomes comparable to the rigid lattice linewidth.

Further evidence of this motion comes from the proton T_1 of benzene (Fig. 5.21(b)) which goes through a minimum at 170 K. Evidently the rotation frequency matches the proton resonance frequency, 23.4 MHz in this case, at this temperature (see Fig. 5.5).

Use of selectively deuterated benzenes allows the intra- and intermolecular contributions to the relaxation to be separated and gives an intramolecular H–H distance of 2.495 ± 0.018 Å, in fine agreement with predictions based on the structure of benzene determined by X-ray crystallography. The activation energy and pre-exponential factor for the spinning of the benzene molecules come out at 15.5 kJ mol^{-1} and 9.1×10^{12} s^{-1}.

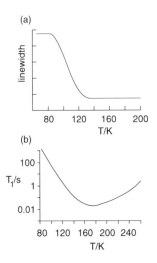

Fig. 5.21 Temperature variation of (a) the ^{1}H linewidth, and (b) the ^{1}H spin–lattice relaxation time of solid benzene. (After E. R. Andrew and R. G. Eades, *Proc. Roy. Soc. (London)*, 1953, **A218**, 537.)

6   Experimental methods

This final chapter deals briefly with NMR instrumentation, before describing a simple pictorial model of magnetic resonance, which is used to discuss two techniques for measuring relaxation times, and two examples of a powerful class of experiments known as two-dimensional NMR.

6.1 Instrumental requirements

The early years of NMR spectroscopy (1945–1970) were dominated by *continuous wave* methods, using a weak, fixed-amplitude radiofrequency field. Spectra were obtained either by keeping the electromagnetic frequency fixed, while slowly sweeping the magnetic field strength, or vice versa, so as to bring spins with different chemical shifts sequentially into resonance, recording the NMR signal all the while. It was essential that the time taken to sweep through each NMR line was *long* compared to its T_1 and T_2, to avoid the distorted signals arising from non-equilibrium spin states. This approach has now been almost completely superceded by *pulse techniques* employing short, intense bursts of radiofrequency radiation. This section gives a very brief description of a pulse NMR spectrometer comprising magnet, radiofrequency transmitter, NMR probe, radiofrequency receiver, and computer. The discussion is restricted to the NMR of liquid samples; solids require somewhat different techniques (see, for example, Harris 1983).

To be suitable for high resolution NMR, the magnet must generate a field that is *strong, homogeneous* and *stable*. There are three principle advantages of using as high a magnetic field as possible: (i) for optimum sensitivity (the strength of the NMR signal increases with the energy-level splitting, eqn 1.12, which is in turn determined by B_0, eqn 1.9); (ii) to maximize the separation of multiplets (chemical shift differences increase linearly with B_0, while spin–spin coupling constants are independent of B_0, so that high fields reduce multiplet overlap); (iii) to minimize strong coupling effects (Section 3.5). An exceedingly *uniform* field is also required for the best possible resolution; any spatial variation in the field experienced by the sample leads to a spread in resonance frequencies and an unwelcome linebroadening (Fig. 6.1). Homogeneity better than one part in 10^9 is required to get 1 Hz linewidths on a high field (e.g. 500 MHz) spectrometer. Finally, the field must not drift by more than one part in 10^9 during the course of an NMR experiment, which may take anything from a few minutes to several days.

These very stringent specifications are only met by superconducting solenoids—coils of resistance-free alloy carrying a persistent current. The alternatives—permanent magnets and electromagnets—are incapable of

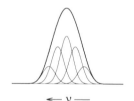

Fig. 6.1 The broadening of an NMR line resulting from spatial inhomogeneity of the static magnetic field B_0. Signals are drawn for *five* regions of the sample experiencing different values of B_0. The amplitude of each contribution is proportional to the number of spins in that region and the bold line represents the lineshape observed for the sample as a whole. In reality there is a continuous, rather than discrete, distribution of B_0 field strengths, and therefore resonance frequencies.

producing fields much in excess of 2.35 T (100 MHz ^{1}H frequency), and are rarely used these days. At the time of writing, the highest field superconducting magnet used for NMR is 17.6 T, giving a ^{1}H frequency of 750 MHz.

Magnetic field *homogeneity* better than one part in 10^9 is achieved by designing and constructing the superconducting solenoid with the utmost care, by using *shim coils*, and by *spinning* the sample. The shims are small coils of wire, placed around the sample, with geometries suitable for generating magnetic field *gradients* of various forms, which cancel the field gradients produced by the solenoid itself. The final improvement in homogeneity is obtained by spinning the sample, typically about 1 ml of liquid in a thin-walled, 5 or 10 mm diameter cylindrical glass tube, at rates of up to 50 Hz. By sweeping the spins through the residual field gradients in this way, the associated linebroadening is removed in much the same way that fast chemical exchange averages chemical shift differences (Section 4.1).

Finally, the desired *stability* is attained by means of a *field-frequency lock*. The B_0 field strength is monitored via the NMR frequency of a reference compound, usually the ^{2}H signal of a deuterated solvent, and any variation in it is used to control the strength of a small supplementary magnetic field which compensates for the drift in the main field.

The most important components of the *radiofrequency transmitter* are shown schematically in Fig. 6.2. A waveform generator produces a continuous voltage oscillating sinusoidally at the desired radiofrequency. This output is chopped into pulses by a switch or 'gate' opened and closed by a pulse generator, which is in turn controlled by the spectrometer computer. The pulses are amplified and sent to the NMR probe.

The probe holds and spins the sample, couples the radiofrequency field to the spins, and picks up the ensuing NMR signal. Its crucial component is a coil of wire or foil placed around the sample, to which the transmitter pulses are applied; the alternating current in the coil generates a magnetic field with the same frequency and phase as the transmitter. The precessing

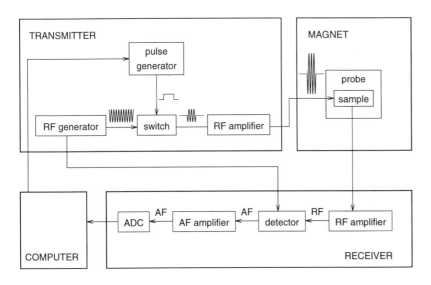

Fig. 6.2 The essential parts of a Fourier transform NMR spectrometer. Abbreviations: RF radiofrequency; AF audiofrequency; ADC analogue-to-digital converter.

nuclear magnetization excited by this radiofrequency field induces in the coil an oscillating voltage, the NMR signal, which is passed to the *receiver*.

After amplification, the NMR signal from the probe is mixed with a reference voltage, usually of the same frequency as the pulses used to excite the spins. This mixing subtracts the reference frequency from the NMR signal to produce an audiofrequency voltage (up to a few kHz), which is amplified further, digitized and processed in the computer.

See Derome (1987) and Freeman (1987) for more information on NMR spectrometers.

6.2 The vector model

The preceding section raises several fundamental questions about the theory and practice of NMR experiments. How does a monochromatic radiofrequency pulse excite the nuclear spins in the sample? Why does this lead to the oscillating magnetization detected by the NMR receiver? How is this signal related to the spectrum? Why bother to do the experiment with radiofrequency *pulses*? The answers, together with insight into the operation of more complex NMR techniques, are conveniently obtained by consideration of a simple *vector model* of magnetic resonance.

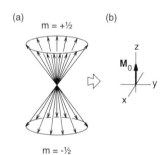

Fig. 6.3 Vector model of a collection of spin-$\frac{1}{2}$ nuclei at thermal equilibrium in a magnetic field along the z axis. (a) The magnetic moments of the individual spins: there is a slight excess of nuclei in the $m = +\frac{1}{2}$ state, which is lower in energy than the $m = -\frac{1}{2}$ state (for nuclides with positive gyromagnetic ratio). At equilibrium, the phases of the individual magnetic moments in the *xy* plane are random. (b) The net magnetic moment of a large number of spins.

Thermal equilibrium

As outlined in Chapter 1, a spin-$\frac{1}{2}$ nucleus in a strong magnetic field (B_0) exists in one of two orientations with equal and opposite projections onto the field direction (the z axis). Depending on its magnetic quantum number ($m = \pm\frac{1}{2}$) each spin contributes $\pm\frac{1}{2}\gamma\hbar$ to the total z magnetization of the sample. At thermal equilibrium, where the population difference is Δn_{eq}, the sample therefore has a net z magnetization

$$M_0 = \tfrac{1}{2}\gamma\hbar\Delta n_{eq} \ . \tag{6.1}$$

In the perpendicular directions x and y, however, the *phases* of the individual nuclear magnetic moments are random, because there is no transverse magnetic field to align them, and their vector sum vanishes (Fig. 6.3). Thus, the total magnetization of the very large number of spins in an NMR sample has magnitude M_0 and is aligned along the positive z axis.

Though made up from the individual *quantum mechanical* nuclear magnetic moments, the total magnetization of the sample conveniently obeys *classical mechanics*. We can therefore forget about quantum mechanics, at least when discussing simple NMR experiments.

The rotating frame

We start by considering a sample consisting of a large number of identical, non-interacting spin-$\frac{1}{2}$ nuclei. The motion of the total magnetization of the spins $\boldsymbol{M}$ in a magnetic field $\boldsymbol{B}$ is shown in Fig. 6.4: $\boldsymbol{M}$ *precesses* around the field direction at an angular frequency (radians per second)

$$\omega = \gamma B \tag{6.2}$$

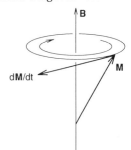

i.e. $\gamma B/2\pi$ revolutions per second. Perhaps not surprisingly, the stronger the magnetic field and the larger the gyromagnetic ratio of the nuclei, the faster the precession. This behaviour is analogous to the motion of the axis of a gyroscope or spinning top in a gravitational field.

In an NMR experiment, $\boldsymbol{B}$ has two components: the strong, static field $\boldsymbol{B}_0$ along the z axis (chemical shifts are ignored for the time being), and a weak radiofrequency field $\boldsymbol{B}_1(t)$ which rotates in the xy plane at a frequency ω_{rf}. The total field (the vector sum of $\boldsymbol{B}_0$ and $\boldsymbol{B}_1(t)$) is thus tilted slightly away from the z axis, and rotates around it at frequency ω_{rf} (Fig. 6.5(a)).

Since it is difficult to envisage the precession of $\boldsymbol{M}$ around a field that is itself moving, we need to simplify the problem. This is done by imagining ourselves to be rotating around the z axis in step with the radiofrequency field, i.e. at frequency ω_{rf}. In this so-called *rotating frame*, the radiofrequency field conveniently appears stationary, say along the x axis. But what happens to the static field? In the *laboratory frame* (i.e. the real world), and in the absence of $\boldsymbol{B}_1$, $\boldsymbol{M}$ should precess around $\boldsymbol{B}_0$ at frequency $\omega_0 = \gamma B_0$. If ω_0 were identical to the radiofrequency ω_{rf}, then $\boldsymbol{M}$, like $\boldsymbol{B}_1$, would appear stationary in the rotating frame; in general, the frequency with which $\boldsymbol{M}$ appears to precess around the z axis, is reduced from ω_0 to:

$$\Omega = \omega_0 - \omega_{rf}. \qquad (6.3)$$

Using eqn 6.2, an effective field ΔB along with the z axis in the rotating frame can be defined as follows:

$$\Delta B = \Omega/\gamma = (\omega_0 - \omega_{rf})/\gamma = B_0 - \omega_{rf}/\gamma \qquad (6.4)$$

whose magnitude is determined by the offset between the precession frequency ω_0 and the radiofrequency ω_{rf}. In the rotating frame, therefore, $\boldsymbol{M}$ precesses around the resultant of $\boldsymbol{B}_1$ and ΔB (Fig. 6.5(b)). Although B_0 is always much greater than B_1, the residual field ΔB can be comparable to B_1 if the transmitter frequency is close to resonance ($\omega_{rf} \approx \omega_0$). Note that ΔB becomes dependent on the chemical shift of the nuclei in question when B_0 is replaced by the field actually experienced by the nuclei, $B_0(1 - \sigma)$, as in eqn 2.2. The precession frequency in the laboratory frame in the absence of $\boldsymbol{B}_1(t)$ then becomes $\gamma B_0(1 - \sigma)$, which is nothing other than the NMR frequency of the spins.

The concept of the rotating frame is exceedingly convenient not only because it removes the unpleasant need to think about time-dependent fields, but also because NMR spectrometers detect offset frequencies ($\Omega = \gamma\Delta B$) rather than the actual resonance frequencies (ω_0), as mentioned in Section 6.1. From now on, our discussion will be based exclusively in the rotating frame.

Radiofrequency pulses

The simplest NMR experiment involves applying a single, short, intense burst of monochromatic radiofrequency radiation to a sample, previously at thermal equilibrium. The transmitter frequency ω_{rf} is set close to the resonance frequency of the spins ω_0, so that the offset field ΔB is small

Fig. 6.4 The motion of a magnetization $\boldsymbol{M}$ in a magnetic field $\boldsymbol{B}$. The rate of change of $\boldsymbol{M}$ ($d\boldsymbol{M}/dt$) is perpendicular to both $\boldsymbol{M}$ and $\boldsymbol{B}$, such that $\boldsymbol{M}$ precesses around $\boldsymbol{B}$ at an angular frequency $\omega = \gamma B$. The magnitude of the magnetization, and the projection of $\boldsymbol{M}$ onto $\boldsymbol{B}$, remain unchanged.

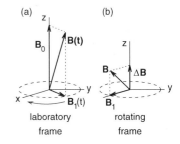

Fig. 6.5 The magnetic fields present in an NMR experiment. (a) In the 'laboratory frame', the net field $\boldsymbol{B}(t)$ is the vector sum of $\boldsymbol{B}_0$, the strong static field along the z axis, and $\boldsymbol{B}_1(t)$, the electromagnetic field rotating around $\boldsymbol{B}_0$ at frequency ω_{rf}. (b) In a coordinate system rotating in step with $\boldsymbol{B}_1(t)$ (the 'rotating frame'), the time-independent effective field $\boldsymbol{B}$ is composed of the offset field ΔB, and the radiofrequency field $\boldsymbol{B}_1$.

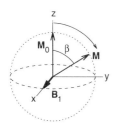

Fig. 6.6 The effect of a radiofrequency pulse on a collection of nuclear spins at equilibrium is to rotate the magnetization away from the *z* axis, around the direction of B_1, through an angle β (eqn 6.5).

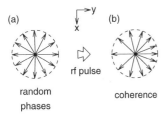

(a)   (b)

random phases   coherence

rf pulse

Fig. 6.7 The effect of a radiofrequency pulse on the magnetic moments of the individual spins in an NMR sample (looking down the *z* axis). Starting from the equilibrium state with random phases (a), a pulse along the *x* axis in the rotating frame causes the spins to 'bunch together' to some extent (b), producing a net *y* magnetization in the sample. This phase-correlation amongst the spins is known as *coherence*.

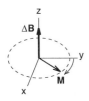

Fig. 6.8 Following a 90° pulse, the magnetization precesses around the *z* axis in the rotating frame at frequency γΔ*B* (ignoring relaxation).

compared to the strength of the radiofrequency field, i.e. $B_1 \gg \Delta B$. Under these conditions the field experienced by the spins in the rotating frame is simply B_1. Choosing the phase of the radiofrequency field such that B_1 lies along the rotating frame *x* axis, the pulse causes *M* to precess in the *yz* plane at angular frequency γB_1, Fig. 6.6. The angle through which the magnetization turns is called the flip angle, β:

$$\beta = \gamma B_1 t_p \qquad (6.5)$$

where t_p is the duration of the pulse in seconds and β is in radians. The most commonly used pulses have 90 or 180° flip angles: a 90° pulse rotates M_0 from the *z* axis to the *y* axis, while a 180° pulse inverts M_0 leaving it along the negative *z* axis. Different flip angles are achieved by appropriate choices of t_p. For typical radiofrequency field strengths, the 90° pulse length is of the order of 10 μs.

A crucial feature of pulsed NMR is the ability to excite *uniformly* and *simultaneously* nuclei with different chemical shifts (i.e. different ΔB). For example, a typical range of ^{1}H resonance frequencies is 4 kHz (10 ppm × 400 MHz); a 90° pulse of strength $\gamma B_1/2\pi \gg 4$ kHz will therefore rotate onto the *y* axis the magnetization vectors of *all* the protons in the sample, irrespective of their resonance frequencies.

The nature of the excitation is a little unusual. A 180° pulse inverts the population difference Δ*n* (remember that the *z* magnetization M_z is proportional to Δ*n*, see eqn 6.1) by interchanging the populations of the $m = \pm\frac{1}{2}$ states. A 90° pulse equalizes the two populations, at the same time converting the equilibrium magnetization entirely into *y* magnetization, M_y. That is, *coherent* electromagnetic radiation induces a *coherence* amongst the spins such that the orientations of the individual magnetic moments in the *xy* plane are no longer random (Fig. 6.7). This seemingly strange behaviour is fundamental to the success of NMR: further details, including the reasonably straightforward quantum mechanical description of coherent excitation, can be found in Goldman (1988), Munowitz (1988), and Slichter (1990).

Free precession and relaxation

The next step is to discover what happens *after* a 90° pulse. When $B_1 = 0$, the only field remaining in the rotating frame is Δ*B*, along the *z* axis. *M* therefore precesses in the *xy* plane (Fig. 6.8) at frequency

$$\Omega = \gamma \Delta B = \gamma B_0 (1 - \sigma) - \omega_{rf} \qquad (6.6)$$

where the chemical shift has been introduced *via* the shielding constant σ. When several types of nucleus are excited, each therefore gives rise to a magnetization that precesses with a frequency characteristic of its chemical shift.

Up to now, we have ignored relaxation (Chapter 5). While this is normally an excellent approximation during the short radiofrequency pulse, we cannot pretend that the spins will not return to equilibrium once the pulse has been switched off.

There are two distinct relaxation processes. First, the recovery of the z magnetization to its equilibrium value, which requires spins to flip so as to re-establish the Boltzmann population difference; and second, the decay to zero of the xy magnetization, by randomization of the phases of the individual spins. The former is simply *spin–lattice relaxation*, which occurs with an exponential time constant T_1 (Section 5.1). More subtly, the latter is *spin–spin relaxation* (time constant T_2). In fact, the loss of phase-coherence between the spins and the linebroadening discussed in Section 5.5 are just different aspects of the same phenomenon. An analogy may help to clarify this point. A well-tuned bell, when struck, produces a single pure note which takes several seconds to die away; once detuned, the bell delivers a short-lived 'clank' comprising a distribution of frequencies. Thus, prolonged ringing goes hand in hand with a narrow band of frequencies, while rapidly damped oscillations correspond to a broad frequency response. So it is in pulse NMR experiments, where slow spin–spin relaxation (i.e. inefficient phase randomization) leads to narrow NMR lines. The random local magnetic fields discussed in Chapter 5 produce small time-dependent variations in the precession frequencies of individual spins which lead to loss of phase-coherence in the sample as a whole.

Figure 6.9 shows the time dependence of M following a 90° pulse. The oscillating, decaying transverse magnetization is detected by the NMR spectrometer via the voltage induced in the receiver coil. This signal, known as the *free induction decay*, is the sum of the individual oscillating voltages from the various nuclei in the sample, each with characteristic offset frequency (i.e. chemical shift and spin–spin couplings), amplitude and T_2. It contains all the information necessary to obtain an NMR spectrum.

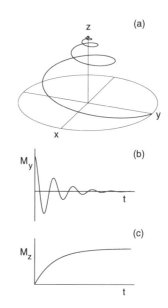

Fig. 6.9 Following a 90° pulse, the magnetization precesses around the z axis and at the same time returns to its equilibrium position along the z axis (a). The transverse components of M decay towards zero with characteristic time T_2, the spin–spin relaxation time (b). The z component grows back to M_0 with time constant T_1, the spin–lattice relaxation time (c).

The NMR spectrum

The final step is to unravel all these oscillating components in the free induction decay to obtain a spectrum of NMR intensity $I(\omega)$ as a function of frequency ω. This is done by means of a *Fourier transform* (FT). In its simplest form the FT of the free induction decay, assumed to be proportional to M_y, is:

$$I(\omega) = \int_0^\infty M_y(t) \cos \omega t \, dt \quad . \tag{6.7}$$

It is easy to see how this works (Fig. 6.10). Suppose the free induction decay contains a component at frequency Ω, $A\cos \Omega t$, where A is the unknown amplitude of the resonance (proportional to the number of spins with offset Ω), and spin–spin relaxation is ignored for the moment. Multiplying this by $\cos \omega t$, as in eqn 6.7, gives $A\cos \Omega t \cos \omega t$, a function that oscillates with positive and negative values, and has *zero* integral over the range $0 \leq t \leq \infty$. That is, the spectral intensity $I(\omega)$ at this particular frequency equals zero. This happens for all values of ω except one: when $\omega = \Omega$ the function to be integrated is $A\cos^2 \Omega t$ which is always positive, with the result that $I(\Omega)$ is proportional to A. Thus Fourier transformation of a free induction decay

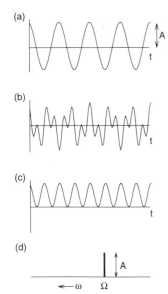

Fig. 6.10 A pictorial representation of the Fourier transform operation (eqn 6.7). (a) A model free induction decay oscillating at frequency Ω, as a function of time t (ignoring relaxation). (b) The product of (a) with $\cos \omega t$ ($\omega \neq \Omega$), which clearly has zero integral over the time interval $0 \leq t \leq \infty$. (c) The product of (a) with $\cos \omega t$ ($\omega = \Omega$), the integral of which is proportional to A, the amplitude of the free induction decay. (d) The NMR spectrum corresponding to (a).

comprising a single component $A\cos \Omega t$ gives a spectrum $I(\omega)$ that is everywhere zero except for a 'stick' of height A at frequency Ω.

As the Fourier transform is a linear operation, i.e.

$$FT[f(t) + g(t)] = FT[f(t)] + FT[g(t)] \tag{6.8}$$

the whole process works however many oscillatory components are present in the free induction decay: for every signal $A_i \cos \Omega_i t$, the spectrum contains a stick of height A_i at frequency Ω_i. Finally, the effect of spin–spin relaxation is to damp the free induction decay exponentially, and so to broaden the sticks in the spectrum $I(\omega)$.

Why bother?

Given that Fourier transform NMR (FT NMR as it is known) requires a considerable investment in equipment—radiofrequency amplifier, pulse programmer, computer, etc.—there must be compelling reasons for preferring it to the older, more straightforward continuous wave experiment. The principal advantage of FT NMR is *sensitivity*. Exciting and detecting the whole spectrum in one 'shot' is much more efficient than sweeping slowly through the spectrum, detecting one resonance at a time. For example, a continuous wave spectrum might take 100–1000 times longer to record than a single free induction decay. The time saved can be used to improve sensitivity, by recording and adding together several hundred or thousand free induction decays. The NMR responses, which are identical in every free induction decay, build up in proportion to the number of signals recorded, N, while the inevitable noise, which varies randomly from one measurement to the next, adds up more slowly, as $N^{1/2}$. The result of coadding N free induction decays is therefore an improvement in the signal-to-noise ratio of $N/N^{1/2} = N^{1/2}$ (Fig. 6.11). For a large number of transients, the improvement in sensitivity can be considerable. Mainly as a result of FT methods, NMR of ^{13}C at natural isotopic abundance is now routine, and the detection of many low sensitivity nuclei has become feasible.

The time-saving aspect of pulse NMR, known as the *multiplex* or *Fellgett advantage*, may be illustrated by considering how one might determine the natural frequency of a bell. One approach would be to set up a frequency synthesizer and loudspeaker, and slowly tune the frequency through the audio range until the bell started to resonate, at which point the frequency could simply be read off from the synthesizer. Clearly this would be a time-consuming and tedious business. An obvious and more efficient method would be simply to strike the bell, and determine the frequency with which it oscillates using a microphone and oscilloscope. Evidently campanologists discovered the merits of pulse excitation over continuous wave methods some time before NMR spectroscopists.

The other major advantage of pulse NMR is that it opens up the possibility of a vast number of new experiments in which the spectroscopist controls the information contained in the free induction decay by using *sequences* of pulses designed to excite the spins in a specific manner. To give just a glimpse of the huge number of different variations on this theme, we start by looking at two simple experiments for determining the relaxation times T_1 and T_2, before moving on to something a bit more sophisticated.

6.3 Relaxation time measurements

Inversion recovery—measurement of T_1

Spin–lattice relaxation times may be measured using the *pulse sequence* $180° - \tau - 90°$ shown in Fig. 6.12. The equilibrium magnetization (a) is inverted by the first pulse, leaving M along the negative z axis (b). During the delay τ, M undergoes partial spin–lattice relaxation (c) to give a z magnetization, $M_z(\tau)$, which the 90° pulse rotates onto the y axis (d). The free induction decay is recorded and Fourier transformed to give a spectrum containing peaks whose NMR intensities $I(\tau)$ are proportional to $M_z(\tau)$. The whole process is repeated for different values of τ so as to map out the recovery of the inverted magnetization (f). Assuming exponential relaxation:

$$M_z(\tau) = M_0[1 - 2\exp(-\tau/T_1)] \qquad (6.9)$$

so that the T_1 of each peak can be obtained by plotting $\ln[I(\infty) - I(\tau)]$ against τ, where $I(\infty)$ is the fully relaxed NMR intensity for $\tau \gg T_1$.

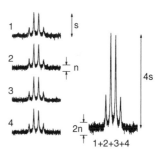

Fig. 6.11 The signal-to-noise ratio of an NMR spectrum can be improved by signal averaging. Coaddition of four independently recorded spectra, each with signal height s and noise standard deviation n gives a spectrum with signal $4s$ and noise $2n$. The signal-to-noise ratio is increased by a factor of two. More generally, the improvement is $N^{1/2}$, where N is the number of spectra added together.

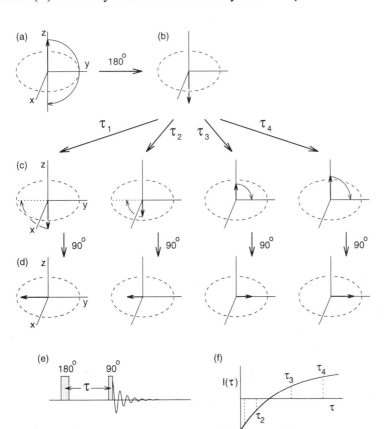

Fig. 6.12 Operation of the inversion recovery experiment for measurement of spin–lattice relaxation times. (a) Equilibrium. (b) After the 180° pulse. (c) After four different delays $\tau_1 < \tau_2 < \tau_3 < \tau_4$. (d) After the 90° pulse. (e) The pulse sequence. (f) the observed NMR signal intensity $I(\tau)$ as a function of the delay τ.

Spin echoes—measurement of T_2

As discussed above, NMR linebroadening and the decay of free induction signals arise from interactions that cause the spins to precess at slightly different frequencies, so destroying their phase-coherence. Two distinct factors are responsible: first, the spin–spin relaxation induced by intra- or intermolecular magnetic fields, and second, the spatial inhomogeneity of the static field B_0. The latter is of no chemical interest but is an inevitable and unfortunate feature of most NMR experiments. As the width of an NMR line depends on both factors, spin–spin relaxation times cannot simply be obtained from linewidths (eqn 5.9). A technique that allows one to separate the two processes is the *spin echo* experiment, shown in Fig. 6.13(a)–(f).

To understand the role played by field inhomogeneities, imagine the sample to be composed of a number of small regions, each of which experiences a different uniform field; spin–spin relaxation is ignored for the time being. After the 90° pulse (b,c), the total magnetization of each region precesses at a slightly different frequency determined by the local magnetic field strength, so that the magnetization vectors fan out in the xy plane (d). This dephasing reduces the transverse magnetization of the sample as a whole, ultimately to zero. After a period τ, the 180° pulse flips the magnetization of each region around the x axis to symmetrical positions in the xy plane (e), whence precession continues for a further time τ. Whatever the precession frequency, and whatever the value of τ, all regions come back into phase perfectly at the end of this second delay (f). This is called a *spin echo*: the dephasing caused by field inhomogeneity is 'refocused' by the 180°

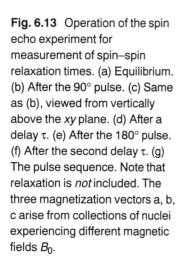

Fig. 6.13 Operation of the spin echo experiment for measurement of spin–spin relaxation times. (a) Equilibrium. (b) After the 90° pulse. (c) Same as (b), viewed from vertically above the xy plane. (d) After a delay τ. (e) After the 180° pulse. (f) After the second delay τ. (g) The pulse sequence. Note that relaxation is *not* included. The three magnetization vectors a, b, c arise from collections of nuclei experiencing different magnetic fields B_0.

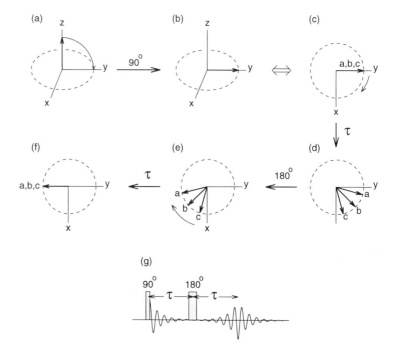

pulse and the NMR signal, which decayed rapidly after the 90° pulse, reappears at 2τ (g). The second half of the echo, which has the same form as the free induction decay, is detected and Fourier transformed to give a spectrum containing NMR lines whose amplitudes are independent of the field inhomogeneity.

Now consider the effect of relaxation on the echo amplitude. Throughout both τ delays, spin–spin relaxation destroys the phase-coherence created by the 90° pulse, and causes the transverse magnetization to decay at a rate T_2^{-1}. This dephasing, which is a product of the fluctuating magnetic fields arising from random molecular motions, is *not* refocused by the 180° pulse. The NMR intensity of each line in the spin echo spectrum is therefore given by

$$I(2\tau) = I(0) \exp(-2\tau/T_2) \ . \tag{6.10}$$

The whole experiment is repeated with different τ delays, and T_2 is obtained from a plot of $\ln[I(2\tau)]$ against τ.

Inversion recovery and spin echoes are two of the simplest NMR experiments. Descriptions of some of the huge number of other techniques are given by Derome (1987), Freeman (1987), and Goldman (1988).

6.4 Two-dimensional NMR

Over the last fifteen years, two-dimensional (2D) methods have revolutionized the practice of NMR spectroscopy; the remaining pages are devoted to an outline of the general principles, and two simple but important examples, of 2D NMR.

The simplest 2D NMR pulse sequence (Fig. 6.14 (g)) comprises two 90° pulses separated by a period t_1 of free precession, and followed by detection of the free induction decay as a function of time t_2. The whole process is repeated with different values of the delay t_1.

Figure 6.14 shows the behaviour of the magnetization of a sample consisting of identical spins, with resonance offset Ω. After the first pulse (b), M precesses in the xy plane at frequency Ω through an angle Ωt_1 (c), at which point the second pulse (d) flips the y component of M onto the z axis, but leaves the x component untouched (a B_1 field along the x axis does not affect M_x). During t_2, this transverse magnetization (e) precesses once again at frequency Ω (f). Thus, the free induction decay has an initial amplitude equal to the x magnetization remaining after the second pulse, i.e. $M_0 \sin \Omega t_1$.

Repetition of the experiment for a range of t_1 values leads to a two-dimensional set of NMR data, as a function of the two time-variables t_1 and t_2, with the signals in t_2 amplitude-modulated as a function of t_1 at frequency Ω. Fourier transformation in both dimensions produces a two-dimensional spectrum as a function of two frequency variables, ω_1 and ω_2, respectively. In this simple example, where the detected magnetization precesses at frequency Ω during t_2, and is amplitude-modulated at the same

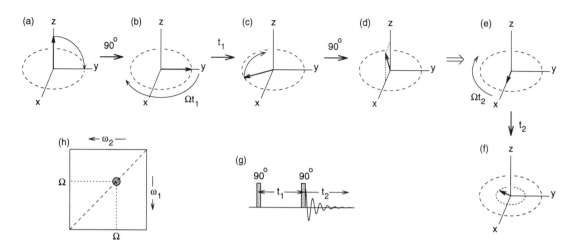

Fig. 6.14 A simple two-dimensional NMR experiment on a sample of nuclei with frequency Ω (in the rotating frame). (a) Equilibrium. (b) After a 90° pulse. (c) After a delay t_1. (d) After a second 90° pulse. (e) Same as (d) with the z magnetization omitted for clarity. (f) After a delay t_2. (g) The pulse sequence. (h) The two-dimensional NMR spectrum.

frequency during t_1, the 2D NMR spectrum contains a single peak at $(\omega_1, \omega_2) = (\Omega, \Omega)$, as shown in Fig. 6.14(h).

Although this example gives some insight into the nature of 2D NMR, it is of course of no practical use: there are much easier ways of measuring the NMR frequency of isolated spins. 2D methods start to become useful when, as is almost always the case, more than one type of nucleus is present. The following paragraphs provide an outline of the workings of two of the simplest and most powerful 2D NMR experiments, which reveal the couplings—scalar or dipolar—between nuclei in a much clearer way than do conventional 'one-dimensional' experiments.

Nuclear Overhauser effect spectroscopy (NOESY)

The pulse sequence for this experiment (pronounced 'nosey') consists of three 90° pulses, the second and third separated by a *fixed* delay τ_m (Fig. 6.15(i)) during which z magnetization is transferred between neighbouring spins *via* the NOE. Consider a pair of protons, I and S, close enough in space to have an appreciable dipolar interaction but, for simplicity, with a negligible scalar coupling. The two rotating frame precession frequencies are Ω_I and Ω_S.

The experiment may be understood by reference to parts (a)–(h) of Fig. 6.15. The equilibrium magnetization of spin I (a) is flipped onto the y axis (b), and precesses at frequency Ω_I (c) for a time t_1. The second pulse flips the y component of this magnetization onto the z axis (d, e). During the delay τ_m, spins I and S undergo cross relaxation and some of the non-equilibrium I-spin z magnetization is transferred to S via the NOE (f). Finally this S-spin z magnetization is rotated onto the y-axis by the third pulse (g), whence it precesses at frequency Ω_S (h) during t_2.

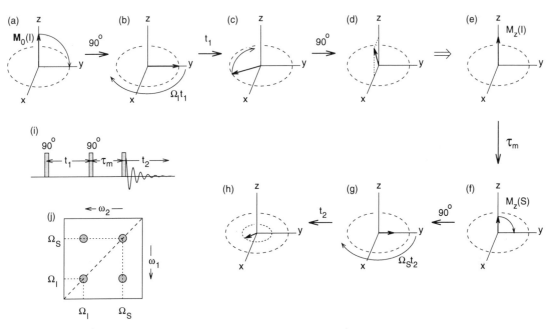

Fig. 6.15 The NOESY experiment on a pair of dipolar-coupled, but not scalar-coupled nuclei I and S. (a) Equilibrium. (b) After a 90° pulse applied to the I spins. (c) After a delay t_1. (d) After a second 90° pulse. (e) Same as (d) with the x magnetization omitted for clarity. (f) After a delay τ_m to allow magnetization to be transferred from I to S by cross relaxation: $M_z(\text{I}) \rightarrow M_z(\text{S})$ (g) After a third 90° pulse. (h) After a delay t_2. (i) The pulse sequence. (h) The two-dimensional NMR spectrum.

The net result of all this is that the free induction decay (frequency Ω_S) is amplitude-modulated at frequency Ω_I. Double Fourier transformation therefore produces a two-dimensional spectrum with a 'cross peak' at $(\omega_1, \omega_2) = (\Omega_I, \Omega_S)$, whose intensity depends on the efficiency with which cross relaxation transfers z magnetization between the two spins during τ_m. Since spin I always has some z magnetization left at the end of τ_m, which gets converted into y magnetization by the final pulse, and precesses at frequency Ω_I during t_2, there is also a 'diagonal' peak at $(\omega_1, \omega_2) = (\Omega_I, \Omega_I)$. By symmetry, excitation of S leads to a cross peak at (Ω_S, Ω_I), and a diagonal peak at (Ω_S, Ω_S), as shown in Fig. 6.15(j). The experiment works in exactly the same way in the presence of an IS scalar coupling. A single NOESY spectrum thus serves to identify all pairs of nuclei with significant NOEs.

Figure 6.16 shows a NOESY spectrum of the protein Fyn whose one-dimensional spectrum appeared at the beginning of Chapter 2. There are several hundred NOE cross peaks, each of which corresponds to a pair of protons separated by less than about 5 Å. As outlined in Section 5.7, the huge amount of information provided by such spectra can be used to deduce three-dimensional structures of molecules.

The NOESY technique can also be used to investigate *chemical exchange*, which like the NOE, is capable of transferring magnetization

Fig. 6.16 600 MHz ^{1}H NOESY
spectrum of the SH3 domain of
the protein Fyn, in D_2O. Each of
the off-diagonal 'blobs' in this
contour plot is an NOE cross
peak. This spectrum was kindly
provided by C. J. Morton.

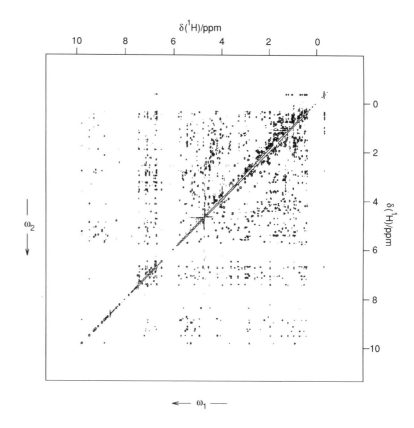

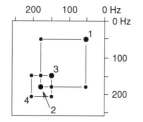

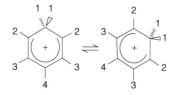

Fig. 6.17 A schematic ^{1}H
two-dimensional spectrum of the
four methyl resonances in the
heptamethylbenzene cation
obtained using the NOESY pulse
sequence. The cross peaks
indicate pairs of protons
undergoing chemical exchange,
in this case a 1,2 methyl shift. A
different mechanism (e.g. a 1,3
shift) would lead to a different
pattern of cross peaks. (After
B. H. Meier and R. R. Ernst,
J. Amer. Chem. Soc., 1979,
101, 6441.)

from one site to another during the delay τ_m. As an example, Fig. 6.17 shows
the spectrum of the heptamethylbenzene cation: the pattern of cross peaks
$(1 \leftrightarrow 2 \leftrightarrow 3 \leftrightarrow 4)$ clearly reveals the existence of an intramolecular rearrangement in which a methyl group undergoes a series of 1,2 shifts.

The advantages of two-dimensional NMR may be seen by comparing
NOESY with its one-dimensional equivalent—a series of ^{1}H$\{^1$H$\}$ NOE
experiments in which each resonance is saturated in turn: (i) crowded spectra are considerably simplified by spreading resonances into a second dimension; (ii) there is no need for selective excitation of individual
resonances, which is difficult or impossible for overlapping lines; (iii) measuring all NOEs simultaneously is more efficient than a series of one-dimensional experiments, in just the same way that FT NMR is superior to
continuous wave NMR.

Correlated spectroscopy (COSY)

In the NOESY experiment, the $90°- \tau_m-90°$ part of the pulse sequence
transfers amplitude-modulated transverse magnetization between pairs of
spins, via cross relaxation or chemical exchange. A similar transfer occurs in
the COSY experiment, by means of a *single* 90° pulse (the second one in the
$90°-t_1-90°-t_2$ pulse sequence, Fig. 6.18(m)) as a result of *scalar* rather than
dipolar couplings. This process is known as *coherence transfer*.

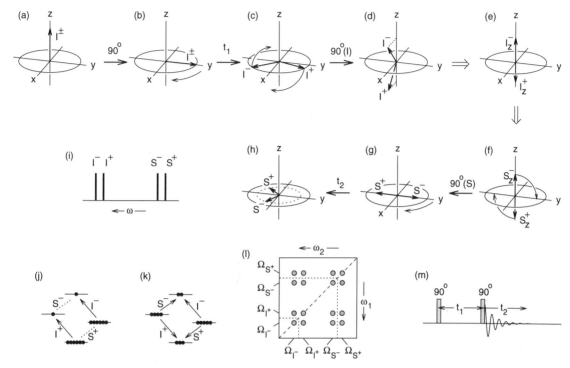

Fig. 6.18 The COSY experiment on a pair of scalar coupled nuclei I and S. (a) Equilibrium. (b) After a 90° pulse applied to the I spins. (c) After a delay t_1, during which the two I-magnetization vectors precess in the xy plane. (d) After a 90° pulse applied to the I-spins. (e) Same as (d) with the x magnetization omitted for clarity. (f) The z magnetization of the S-spins, corresponding to the non-equilibrium state in (e). (g) After a 90° pulse applied to the S-spins. (h) After a delay t_2, during which the two S-magnetization vectors precess in the xy plane. (i) The NMR spectrum of I and S, showing the four NMR transitions $I^{\pm}$ and $S^{\pm}$. (j) Schematic populations of energy levels at the start of the experiment. (k) Schematic populations just after the 90°(I) pulse, corresponding to the z magnetizations shown in (e) and (f). (l) The two-dimensional NMR spectrum. (m) The pulse sequence.

Consider a pair of weakly coupled protons I and S, with a spin–spin coupling constant J. The four allowed ^{1}H NMR transitions are labelled $I^{\pm}$ and $S^{\pm}$, and have resonance offsets $\Omega_{I\pm}$ and $\Omega_{S\pm}$, such that $\Omega_{I-} - \Omega_{I+} = \Omega_{S-} - \Omega_{S+} = 2\pi J$, as indicated in Fig. 6.18(i). To simplify the argument a little, we suppose that both of the S transitions (S^+ and S^-) are initially *saturated*, as indicated on the energy level diagram in Fig. 6.18(j).

The operation of COSY may be understood by reference to parts (a)–(h) of Fig. 6.18. Initially both I^+ and I^- magnetizations lie along the z axis (a); the first 90° pulse flips both onto the y axis (b), whence they precess at their respective offset frequencies Ω_{I+} and Ω_{I-} for a time t_1 (c). The action of the second 90° pulse is most easily seen by considering first its effect on spin I. As in NOESY the transverse magnetization is rotated from the xy plane into the xz plane (d), to give amplitude-modulated z magnetizations I_z^+ and I_z^- (e), and therefore amplitude-modulated populations of the four spin states as shown on the energy level diagram (k). Because the spins are coupled,

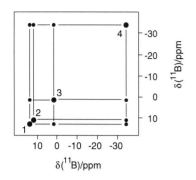

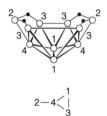

Fig. 6.19 Schematic ^{11}B COSY spectrum of $B_{10}H_{14}$. In the structure, the small solid circles represent bridging hydrogens; the ten terminal hydrogens (one per boron) are not shown. The four types of boron atom give rise to four cross peaks and hence four 'connectivities', shown below the structure. (After T. L. Venable, W. C. Hutton, and R. N. Grimes, *J. Amer. Chem. Soc.*, 1984, **106**, 29.)

such a disturbance of the population differences of the I spins corresponds automatically to changes in the population differences, and therefore z magnetizations, S_z^+ and S_z^-, of S (f, k). The effect of the second 90° pulse on S is simply to flip $S_z^{\pm}$ onto the y axis (g), giving a free induction decay as the S spins precess at frequencies $\Omega_{S\pm}$ (h). In short, the second 90° pulse transfers amplitude-modulated transverse magnetization from I to S and gives cross peaks at $(\omega_1, \omega_2) = (\Omega_{I+}, \Omega_{S+})$, $(\Omega_{I+}, \Omega_{S-})$, $(\Omega_{I-}, \Omega_{S+})$ and $(\Omega_{I-}, \Omega_{S-})$, Fig. 6.18(l). As in NOESY, diagonal peaks arise from the portion of the I-spin magnetization that is not transferred to S. Removal of the restriction that the S spins are initially saturated, leads by symmetry to cross and diagonal peaks with $\omega_1 = \Omega_{S\pm}$, and $\omega_2 = \Omega_{I\pm}$ and $\Omega_{S\pm}$ (Fig. 6.18(l)). Inspection of Fig. 6.18 shows that the criterion for observing IS cross peaks in a COSY spectrum is that the scalar coupling J_{IS} must be large enough to allow the two magnetization vectors of each spin to get significantly out of phase during t_1.

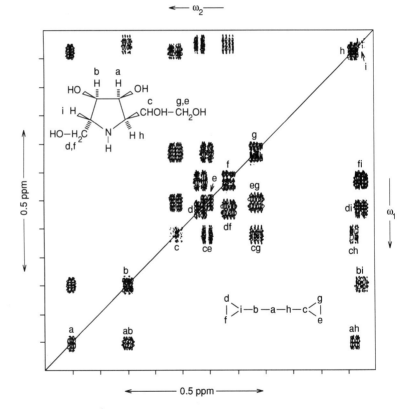

Fig. 6.20 400 MHz ^{1}H COSY spectrum of the ring compound shown (in D_2O). The CH protons are labelled a–i; the labile OH and NH protons are replaced by deuterons in D_2O and do not appear in the spectrum. The structure within the cross and diagonal peaks arises from partially resolved spin–spin coupling. The inset shows the network of scalar couplings revealed by the pattern of cross peaks. This spectrum was kindly provided by M. R. Wormald and G. W. J. Fleet.

The object of a COSY experiment is to discover the network of spin–spin couplings in a molecule, and so to arrive at spectral assignments—i.e. to determine which resonance corresponds to which nucleus. This process is illustrated by the following simple examples.

Figure 6.19 shows the ^{11}B COSY spectrum of the borane $B_{10}H_{14}$. Cross peaks are only observed between directly bonded borons ($4 \leftrightarrow 1,2,3$ and $1 \leftrightarrow 3$) giving immediately an unambiguous assignment of peaks 2 and 4. Resonances 1 and 3, which arise, respectively, from two and four boron atoms, are easily distinguished by their 1:2 intensity ratio in a one-dimensional spectrum.

A slightly more complicated COSY spectrum is shown in Fig. 6.20. Nine CH protons (*a–i*) give rise to ten cross peaks; all two-bond and three-bond spin–spin couplings are observed. Once again the pattern of scalar couplings is trivially deduced. The ambiguity in the assignment, arising from the two-fold symmetry of the connectivity pattern, is easily resolved by chemical shift arguments or, more satisfactorily, by the presence of NOEs from peak *h* to peaks *d* and *f*, and the absence of NOEs from *b* to *g* and *e*.

COSY and NOESY are but two examples of an enormous range of powerful 2D NMR experiments available to the chemist. A myriad different techniques exist, some highly sophisticated, for probing the secrets of molecular structure and motion (Ernst *et al.* 1987). There are even three- and four-dimensional NMR methods for studying ever larger and more complex molecules.

Bibliography

Introductory NMR textbooks include:

Bovey, F. A. (1988). *Nuclear magnetic resonance spectroscopy* (2nd edn). Academic Press, San Diego.
Carrington, A. and McLachlan, A. D. (1967). *Introduction to magnetic resonance*. Harper and Row, New York.
Freeman, R. (1987). *A handbook of nuclear magnetic resonance*. Longman, Harlow.
Günther, H (1980). *NMR spectroscopy*. Wiley, Chichester.
Harris, R. K. (1983). *Nuclear magnetic resonance spectroscopy*. Pitman, London.
McLauchlan, K. A. (1972). *Magnetic resonance*. Clarendon Press, Oxford.
Pople, J. A., Schneider, W. G. and Bernstein, H. J. (1959). *High-resolution nuclear magnetic resonance*. McGraw-Hill, New York.
Wehrli, F. W., Marchand, A. P. and Wehrli, S. (1988). *Interpretation of carbon-13 NMR spectra* (2nd edn). Wiley, Chichester.

Other books referred to in the text:

Atkins, P. W. (1983). *Molecular quantum mechanics* (2nd edn). Oxford University Press.
Derome, A. E. (1987). *Modern NMR techniques for chemistry research*. Pergamon, Oxford.
Ernst, R. R., Bodenhausen, G. and Wokaun, A. (1987). *Principles of nuclear magnetic resonance in one and two dimensions*. Clarendon Press, Oxford.
Goldman, M. (1988). *Quantum description of high-resolution NMR in liquids*. Clarendon Press, Oxford.
Homans, S. W. (1989). *A dictionary of concepts in NMR*. Clarendon Press, Oxford.
Kemp, W. (1986). *NMR in chemistry*. Macmillan, Basingstoke.
Munowitz, M. (1988) *Coherence and NMR*. Wiley, New York.
Neuhaus, D. and Williamson, M. (1989). *The nuclear Overhauser effect in structural and conformational analysis*. VCH Publishers, New York.
Sanders, J. K. M. and Hunter, B. K. (1993). *Modern NMR spectroscopy* (2nd edn). Oxford University Press.
Slichter, C. P. (1990). *Principles of magnetic resonance* (3rd edn). Springer-Verlag, Berlin.
Williams, D. H. and Fleming, I. (1989). *Spectroscopic methods in organic chemistry* (4th edn). McGraw-Hill, London.
Wüthrich, K. (1986). *NMR of proteins and nucleic acids*. Wiley, New York.

Index